Purposeful Engineering Economics

Ronald A. Chadderton Ph.D.

Purposeful Engineering Economics

 Springer

Ronald A. Chadderton
Villanova University
Villanova, PA
USA

ISBN 978-3-319-37535-9   ISBN 978-3-319-18848-5   (eBook)
DOI 10.1007/978-3-319-18848-5

Springer Cham Heidelberg New York Dordrecht London
© Springer International Publishing Switzerland 2015
Softcover reprint of the hardcover 1st edition 2015

Printed on acid-free paper

Springer International Publishing AG Switzerland is part of Springer Science+Business Media
(www.springer.com)

This book is dedicated to my wonderful wife, Donna, whose continuing love and encouragement have made work possible and life complete.

Preface

"Remember the Austrians" or "What Does Catallactics Have to Do with Civil Engineering?"

The title, "Purposeful Engineering Economics," was chosen to emphasize the importance of the intentions behind any economic action. To explain this idea adequately requires investigation well beyond engineering economics, per se. In fact, much of the narrative will range over a wide subject area to draw from works in sociology, philosophy, political economy, history, and even religion for background materials. I am not aware of any previous attempt to "explain" engineering economy this way.

This book is intended primarily for instructors of engineering economy classes, particularly those for civil engineering students. It is assumed that the instructor will be well versed in the methods of engineering economy but is interested to include the wider perspective of catallactics in the course. This book will suggest ways to do this; it is not intended to replace a basic engineering economy textbook.

Although I am not an economist, I created and taught a course on economics and risk for almost 20 years. For many years before that, I maintained an abiding interest in the literature of the Austrian school economists, obviously not a common interest for engineers. Several private foundations provided enough support for the development and continued revision of the CEE course known as "Economy and Risk" as well as several associated case studies. This decade-long relationship produced numerous published articles on subjects as varied as water quality control programs, the transcontinental railroads, and the course content itself. The latest presentation of this material was at an international meeting concerned with water resources and other environmental issues.

I had planned to include a section titled "It All Started with Ayn Rand," but that statement is not really true either chronologically or personally. So, when I recently saw a movie called "Remember the Titans," I promised my class that I would use "Remember the Austrians" as a title in the book, instead.

Later, I realized that this title is more than just a play on words because there seems to be a more meaningful similarity. If you remember the movie, it is about bringing together disparate groups to reach a goal. That is also what Austrian school economists worry about—the coordination of individual efforts based on subjective values so that all participants can remove their respective "felt uneasiness" by purposeful action; they feel better after the action. This coordination problem will have a prominent place in this narrative.

As for Ayn Rand, long ago a friend gave me a copy of "The Fountainhead" and said: "You're an engineer, you'll enjoy this." Eventually, I read my way through her novels and nonfiction. Buried in the reference section of a collection of essays called "Capitalism, the Unknown Ideal," there is a citation of Ludwig von Mises' treatise "Human Action". (I always say that I am probably one of the only two engineers who have ever read that 900-page book, and I have done it twice!) That was the book that introduced the Austrian School of economics to me. I subsequently read my way through a lot of that literature by Mises, Hayek, and others, eventually finding current adherents such as Sowell and Williams, among others. Even now, as this is written, I am finding additional adherents to the more modern, American branch of the Austrian tradition, especially O'Driscoll, Rizzo, Kirzner, Lachman, and Vaughn.

Mises explained the situation this way: "The cornerstone of the Austrian school is the subjective value marginal utility theory. This theory traces all economic phenomena, simple and complex, to the actions of individuals, each undertaken as a result of personal subjective values."

The second half of the title, "What Does Catallactics Have to Do With Civil Engineering?" is a question I have often been asked by friends, faculty members, and manuscript reviewers. Other similar questions I am asked include the following: What are you talking about? Why is this important? What does it have to do with engineering? What does it add to an engineer's education? One reason for this book is to convince at least some readers that ideas such as subjective value and marginal utility do have a significant relationship to civil engineering.

My interest in catallactics did not really begin with "The Fountainhead." Several personal experiences contributed to the desire to write this book. The first incident that I can remember was when I was a 17-year-old high school student. Our "Problems of Democracy" teacher made an indelible impression when he said in class that "even if the tax rate was 99 %, he would continue to work to get the last 1 %." Now, even a 17-year-old who has never heard of economics will realize that his statement just does not sound right, but would not know why. One objective of my course has been to help college students understand the "why" as an individual subjective value decision.

Another incident occurred only a few years ago when a recent graduate of our undergraduate engineering program wrote to tell me that she had been in a meeting with the supervisors in her engineering firm and she was the only one who knew what "Present Value" meant. This was discouraging because they were apparently trying to work out a cash flow analysis for an engineering project.

A final "tipping point," to use that overused idea, has been the misquoting, or even ignoring, of the US Constitution by several high-ranking members of our

Congress. Two recent incidents stand out: one referring to the "health and welfare clause" and the other crediting the 14th Amendment [1868] to the Founders. These are among the reasons that I give each of my students a pocket copy of the US Constitution, like the one that Peter Jennings carried around, as part of my courses. Another reason is that the economy and risk class has a component that addresses "current issues," and among the issues considered have been various governmental activities and Constitutional debates surrounding them.

Justification for incorporating catallactics in engineering education comes from a recent National Academy of Engineering report, "Educating the Engineer of 2020" which stated: "Engineers must learn…basic micro-economics, the setting of prices, the determinants of market value, and so forth." Because "…the best engineering solutions can emerge only in the context of market prices and market forces." Therefore, "…engineers need market prices, not black-and-white regulations, to make correct, 'unwasteful,' economic decisions, and engineers should inject themselves forcefully into this very public debate."

My argument for including the "Economy and Risk" course in an undergraduate curriculum has been that to work for the public good, civil engineers must favor economic efficiency and optimum allocation of scarce resources and that resources, including fear, are always in scarce supply. To accomplish these objectives, the engineer should have an appreciation for the subjects of economic action as best defined by the Austrian economists. They really do not need to know macroeconomics of any variety. My view, supported by Austrian theorists, is that macroeconomics and "econometrics" are fallacious.

This book is organized to follow a particular engineering economy textbook (Thuesen & Fabrycky) used in my classes, although an instructor could easily use a different text. It gives suggestions for applications of Austrian economic principles to several topics of engineering economy. It also includes a section describing basic elements of Austrian economic theory and references for further study. An example set of learning objectives for a course incorporating these principles is appended.

The main objectives of my course are to study cash flow (engineering economy) supplemented with economic fundamentals (e.g., supply and demand analysis) and risk evaluations. The course is informed throughout, both explicitly and implicitly, by references to Austrian school philosophy best represented by the work of Hayek and Mises and including others such as Rothbard and even Menger, himself.

The course is intended to be more than simply cash flow; four elements were interwoven to construct it: fundamental economics (catallactics, a non-mathematical branch of the social sciences); engineering economy (cash flow); risk and uncertainty in decision making; and current events (political economy).

Work in the class should progress from the ideal world (no inflation; no taxes) to the study of the same cases including effects of depreciation, taxes, and inflation. Several example cases are investigated for "sustainability" concerns as well as the effects of government policies on project acceptability.

To return to the question: Why is this manuscript titled "Purposeful Engineering Economics"? My dictionary defines "purposeful" as "having a purpose; intentional" and "purpose" as "a result or effect that is intended or desired."

So, the "intent," herein, is to present a supplement to the traditional engineering economics textbook that elaborates the *consciousness* behind the mathematics of the decision process. Such *consciousness* is a capacity of an individual; as such its content is unknowable to anyone except that individual. Rothbard said that individuals do not behave like billiard balls; their actions do not follow physical laws, they are purposeful, mind-driven.

Acting individuals are attempting to achieve their particular goals in a world filled by uncertainty. Their actions cannot be predicted; to make such a claim is, according to Hayek, the "fatal conceit" of social engineers and planners. According to O'Driscoll and Rizzo, actions occurring through "real time" encounter constantly changing information as actors discover new knowledge amid the uncertainty. No one, not even the acting individual, can foresee the future state of knowledge. Hayek called the attempt to do so the "pretence of knowledge" in his Nobel Prize acceptance lecture.

It has been my intention, both in class and in this book, to give engineering students an appreciation of these Austrian insights so that they can take a more critical view of accepted knowledge. It is my hope that readers of this slim volume will also find something of value.

Acknowledgments

The collection, organization, and presentation of this material have taken a very long time. Many individuals have contributed to it, most without being aware of doing so. Hundreds of students have attended my classes, answered survey questions, or offered constructive criticism.

For many years, I have been fortunate to have been the Edward A. Daylor Endowed Professor at Villanova University, a position that provided time and resources for independent scholarship, much of which has led to this project. The Carthage and Sarah Scaife Foundations also provided support for the development and publication of course materials, particularly several case studies.

Several reviewers, unknown to me, have offered suggestions for improvements to the text. While I have not always followed them, these suggestions have made the project more complete.

Four individuals at Springer have provided invaluable assistance, namely Michael Luby, Merry Stuber, Brian Halm, and Prasanna Kumar N. The manuscript would have been neither completed nor published without their efforts.

My wife, Donna, consistently supported the development of the manuscript, read several drafts, and offered many insightful suggestions that made it much better than it would have been. Without her encouragement, there would have been no manuscript.

Thus, the work you see had many contributors, but any errors of fact or interpretation are mine. I only hope that some instructors will be encouraged to incorporate Austrian concepts in their future classes.

Villanova, PA, USA   Ronald A. Chadderton
February 2015

Contents

Chapter 1
Introduction

This book is the result of many years of two parallel activities; studying the so-called "Austrian" school of economics and teaching classes in engineering economy and risk. Most readers will not be familiar with the terms "Austrian Economic Theory" or "Austrian Economics." This section will explain these terms and describe my method of incorporating some Austrian insights into an undergraduate engineering economy course that was created specifically for my department.

The general objective of my engineering economy and risk class has been to incorporate enough theory to inform students that economics is not simply numerical calculation. This establishes a contradiction because the mathematics of engineering economy, while straightforward, appears in conflict with the purely conceptual nature of Austrian economics. However, explaining the contradiction allows the inclusion of social policy and current events, an increasingly significant component of current engineering education required for accreditation.

My intention is not to present a complete explanation and justification of Austrian economic theory. The references listed at the end of each chapter will include sufficient resources for an interested reader to delve into the extensive literature. A very brief introduction to the most important elements of Austrian theory for my purposes will be given here. Appendix A includes a more extensive introduction to elements of the Austrian analysis.

To establish context for my class and the following discussion, consider some basic elements of economic theory (Seldon 2004):

- Production takes place for consumption.
- Value is measured at the margin, not as an average.
- The cost of a commodity or service is the commodity or service given up.
- Demand and supply of anything vary with price.
- Without price signals, co-operation is impossible.

These ideas will be central to the topics of this book. Their origins will be included in the Reference sections.

© Springer International Publishing Switzerland 2015
R.A. Chadderton, *Purposeful Engineering Economics*,
DOI 10.1007/978-3-319-18848-5_1

1.1 The Basic Problem: "The Economics of Time and Ignorance"

The fundamental problem to be solved is to allow peaceful, voluntary cooperation of acting individuals to remove their respective "felt uneasiness" so that all parties feel more satisfied with their situation, or less "uneasy". Unfortunately, this process must occur in a world of inadequate knowledge. Errors will occur. The question is how best to minimize error and reach satisfactory outcomes. The Austrian analysis describes methods that have the greatest potential to realize success.

I am using the title of the O'Driscoll & Rizzo book (*The Economics of Time and Ignorance*) to describe the state of affairs. We engineers want to believe that our analyses use hard data as constants in our equations to predict answers. We believe that if we use current prices we can make accurate predictions about the future results of our plans. The Austrian analysis contradicts these assumptions.

Economic data, prices for example, are historic information and not predictors of future conditions. Much common knowledge, common sense, will prove to be wrong. Students will be expected to question every assumption. Austrian insights will be used to help explain reasons for these unexpected contradictions.

We continually hear repeated, often from the highest levels, the very fallacies contradicted long ago by the "Austrians." For example, an eminent economist recently claimed that Hurricane Sandy stimulated the economy. I can remember a high-ranking member of Congress saying the same thing about World War II; it was good for the economy; it created many jobs; etc. To continue to contradict these fallacies has become a burdensome obligation, but there are several eminent economists who can be cited to support dissenting views.

Great reliance will be made on the analyses of the conditions of "time and ignorance" presented by O'Driscoll and Rizzo (1996). Menger (1994) will tell us about the origin of money. Hayek (2009) will explain effects of inflation on price relationships. Hazlitt (1996) will show us how to look beyond what is seen in the short term to consider the long-term effects of public policies. Bastiat (1996) will reduce to absurdity situations analogous to current day pronouncements of high-level officials.

On a more practical side, we have seen Fundamentals of Engineering exam scores in engineering economy topics improved for our students since this course has been offered. And, procedures used for engineering project analysis can be demonstrated with examples relevant to personal finance to show students an immediate value for the course.

In a general sense, we must also appreciate the limitations of our knowledge. And, we must insist on proper use of our language. Both of these topics are considered next in greater detail.

1.2   Knowledge: How Much Do We Know; How Confident Should We Be?

A 1973 PBS series called "The Ascent of Man" included a program titled "Knowledge or Certainty." The narrator, Jacob Bronowski, described a multitude of increasingly precise instruments to make observations of matter. This sequence culminated with what he labeled "the crucial paradox of knowledge," that no matter how fine an observation might be, the object will always appear "fuzzy"; it is as "uncertain as ever."

Bronowski (1973) explained Gauss's invention of the normal, bell curve of probability to represent values of repeated observations as statistical summary by mean (average value) and variance (spread of values). He concluded that "errors are inextricably bound up with the nature of human knowledge."

Bronowski renamed the "Principle of Uncertainty" (Heisenberg) the more appropriate "Principle of Tolerance" because "All knowledge, all information between human beings can only be exchanged within a play of tolerance. And, that is true whether the exchange is in science, or in literature, or in religion, or in politics, or even any form of thought that aspires to dogma." He contrasted this tolerance in science, the knowledge that we can know even when fallible, with a despotic belief in absolute certainty, exemplified by Nazi Germany. He concluded that "We have to cure ourselves of the itch for absolute knowledge and power."

In a 1944 book, *Omnipotent Government*, Mises maintained that conflict and war (specifically World War II) was inevitable because of Nazi Germany's interference with business and free exchange. The more totalitarian (interventionist) a government became the more damage it would do to its economy resulting in depression, unemployment, inflation, rising prices, and even war.

Hayek's 1988 book, *The Fatal Conceit*, argued that "the fatal conceit" is the idea that "man is able to shape the world around him according to his wishes"; or, the impossibility of centralized planning (even if well-intentioned) because of the problem of diffused information. No one can have access to the information (data) needed to plan an economy. Hayek addressed this problem in the essay, *The Use of Knowledge in Society* (1977), and in his Nobel Prize acceptance speech, *The Pretence of Knowledge* (1974).

This pretence of knowledge, as described by Hayek, comes from assuming that the methods of the physical sciences, which have achieved such great success, applied in "economics and other disciplines that deal with essentially complex phenomena." He concluded the speech by describing a "fatal striving to control society—a striving which makes him [the "planner," the "social engineer"] not only a tyrant over his fellows, but which may well make him the destroyer of a civilization which no brain has designed but which has grown from the free efforts of millions of individuals." Hayek's *The Constitution of Liberty* (1960) explained the evolution of organizations of which the participants have been unaware, such as languages or economic systems.

A section called "Case Probability" in Mises' *Human Action* (1963) contrasted "class probability" with "case probability." Essentially, Mises maintained that an economic decision is a unique situation, not repeatable in its details, and therefore "not open to any kind of numerical evaluation." In other words, an assignment of probability to the outcome of a future economic action according to a kind of relative frequency of past events is invalid and would prove wrong. He presented a similar argument in *Theory and History* (1957) in which he contrasted historical information with the economic data needed for prospective economic calculation.

O'Driscoll and Rizzo, *The Economics of Time and Ignorance* (1996), noted that "[r]eal time, ignorance, and genuine uncertainty…have significant implications for economic policy." As actions move ahead in "real time" not only is time irreversible, but the data are constantly changing.

We mean by data in this case not only price information but also personal preferences, as preferences would change with changing price information. We can envision a sequence of change in the natural state of an economy; "real time" causes "experience" which yields "new knowledge" that leads to adjusted plans and actions. A consequence of the passage of "real time" is creative evolution causing "genuine uncertainty" and "true surprise" (all terms from O'Driscoll and Rizzo 1996).

The result of this analysis is that one cannot know his own future knowledge and cannot predict his own future decisions; therefore, even self-prediction is impossible. Would not collective prediction by a planner or social engineer be even more likely to be wrong or even disastrous?

If the social engineer could accumulate all of the dispersed, individualized information needed to start his plan, as the plan progressed and the data changed, the plan would become ever more removed from a correct or even its intended trajectory. Thus, because of this inseparability of time and change, economic processes are fundamentally indeterminate. However, in an unplanned society, one could expect a multitude of proposed responses to each "surprise"; many would be wrong, but there would be a significant chance that some would be appropriate. A recent presentation of this argument can be found in Harford's *Adapt; Why Success Always Starts with Failure* (2011).

Many engineers study planning in their courses. Therefore, they are more likely to be receptive to "planning" than the general populace. They should have a better understanding of the likelihood of failure of "planning." Elements of Austrian theory in their education can, at least, cause them to pause to question social planning.

This uncertainty is not risk that can be calculated from known probabilities. The probabilities of future events are unknown; even if they were known at one moment they would be changed by new knowledge a moment later.

An appreciation of this situation should negate macroeconomic analysis and econometrics; instead, we have the macro-driven "Crisis of 2013." The Austrian school of economics has more to say about our situation than any other school. Teachings of the Austrians are dispersed throughout this narrative to expand students' perspective.

1.3 Notes on the "Confusion" of the Language

Just as mathematical equations have meaning behind the symbols, so words must have specific meaning to be useful. Terms must be carefully defined and must be used consistently to avoid confusion.

For several years I have told my classes that words, at least, "should have meaning." My concern goes beyond misuse of the language, in print and on the screen, to the more serious issue of the purposeful corruption of language.

Words must have generally understood meanings for people to be able to communicate. When language is confused, activity cannot be described or coordinated. Communication becomes impossible. Lane (1984) listed numerous meanings for the word "democracy," among them socialism, communism, dictatorships, anarchists, etc. In short, anything or everything, including Bismarck's welfare state could be labeled "democracy."

Can anyone honestly say that such confusion has not occurred? For example, there has been a complete inversion of meaning of the term "liberal" from its original reference to the philosophy of "freedom" to the modern reference to the philosophy of "socialism" or the "welfare state" (Mises 2005). As early as 1941, Mises (2011) wrote that "the old [classical] liberalism has even lost its name."

Both Lane and Mises were writing during World War II, the New Deal era in the USA. One could say that that was an era of expansive planning by an interventionist state. The general, federal, government was expanding beyond its constitutional bounds and assuming functions beyond its legitimate role but upon which the citizens would become dependent. At the same time, government acted as if it knew what its citizens should do better than they themselves (e.g., massive mandatory "retirement" and "insurance" programs) and driving itself into deepening debt. More recently, Hayek maintained that the term "liberty" could be used to support measures which would destroy individual liberty and that the use of collective power substituted for individual liberty resulted in the demand to redistribute wealth (Hayek 2011).

I once gave members at a faculty meeting pocket copies of *The Declaration of Independence and the Constitution of the United States of America* (Cato Institute 2002) and explained that I had passed out hundreds of copies to students because I thought we were in the midst of the greatest Constitutional crisis since the Civil War [1860s]. To say that this was treated with scorn or disbelief would be an understatement! The extent of this crisis should be obvious. I continue to give copies to my students because I consider it to be one of the primary source documents for a course in economics.

One should not assume that there was no warning about these expansive developments. Prominent among those who sounded warning were the economists of the Austrian School. Because of the immediate relevance but generally unknown content of Austrian economics, this narrative will attempt to integrate elements of these teachings into engineering economics.

What is "unique" about Austrian economics and how does it relate to "engineering economy"? Why is this important, and why should anyone (least of all, students) care? Part of the answer can be found in the confusion in Washington, D.C. The Constitution is about freedom, not entitlement (Arnn 2012; McClanahan 2012); Austrian economics is about freedom, also. The two belong together. The University is a place where "inquiry is pushed forward,…and error exposed…." (Newman 1987). Unfortunately, almost in parallel with the expansion of the general government, universities have expanded their roles, voluntarily. As Barzun (1993) noted, "…in accepting the role of public service center and the moneys that go with it, the university has made itself answerable to eight publics, whose acts and words often turn them into foes….The eight publics are: parents, students, alumni, donors, foundations, neighbors, governments, and newspapers. There may be nine if there exists a general public not already distributed among the eight."

Curtler (2001) maintained that the purpose of higher education is to help students (citizens) become free, exploring connections of freedom, education, and citizenship. These are connections explored in my course.

Institutions in a democratic society are responsible to prepare citizens for "reasonable action in the political sphere." The goal of higher education and the requirements of a free society overlap. But, according to Curtler (2001), "…the single, clear role of the academy in our society has become confused…[due to]… increasing ineffectiveness of the family and churches as viable social institutions. As a consequence of this degeneration, our schools are now expected to solve every problem, to be all things to all people, and this exacts promises from the schools that they cannot possibly hope to deliver." This assessment clearly agrees with Barzun (1993), schools (especially higher education) make too many promises that they cannot keep.

Curtler (2001) identified three major causes of confusion of purpose in higher education: over-administration, over-specialization, and lack of concern with [real] education. The overall result has been to "weaken and fractionalize the central purpose of education."

Curtler (2001) contradicted five myths about purposeful education, namely "education is job training," "education is schooling," "education is information," "education is enculturation," and "education is a market place, students are consumers of a product." He concluded that "the academy must once again focus attention on the ideal of liberal education as the attainment of positive (internal) human freedom." And, to do so requires that "all other demands that have been placed on it must stand and wait their turn."

Including an Austrian perspective on engineering economics is consistent with both Barzun and Curtler. It is not presented as the only perspective, but forces the students to consider ideas that they have not encountered elsewhere. In my terms, it makes them "question the unquestionable."

The University is still a place where Austrian style economics and Constitutional freedom can, at least, be mentioned together. As I remind my students, we live in a country where we can still read our founding documents. I urge them to believe that the Founders wrote what they meant and meant what they wrote.

The two central elements of Austrian economics are time and ignorance. This means the passage of "real time," the evolution of knowledge, and the real uncertainty of an unknowable future. In the face of time value over real time in the presence of an unknowable future, should our leaders not exhibit more humility and less arrogance? Should they not exhibit restraint instead of pushing everyone into possibly ill-conceived and unwanted massive projects, such as welfare entitlement programs? Should not these questions be considered by students, the future citizens of this country?

For all these apparently interrelated reasons, I attempt to integrate some elements of Austrian economic theory, some of the original intent of the Constitutional Founders, and some incidents from history into my engineering economics courses. At the least, students should be asked to question the status quo while it is still possible to do so.

To conclude this section, consider the following selection from Mises' *Human Action* that seems to be a good reason for engineering students to learn some Austrian economic theory: "The art of engineering can establish how a bridge must be built in order to span a river at a given point and to carry definite loads. But it cannot answer the question whether or not the construction of such a bridge would withdraw material factors of production and labor from an employment in which they could satisfy needs more urgently felt. It cannot tell whether or not the bridge should be built at all, where it should be built, what capacity for bearing burdens it should have, and which of the many possibilities for its construction should be chosen." (Mises 1996)

References

Arnn L (2012) The founders' key. Thomas Nelson, Nashville

Barzun J (1993) The american university. Chicago

Bastiat F (1996) Economic sophisms. Foundation for Economic Education, Irvington-on-Hudson, NY

Bronowski J (1973) The ascent of man. Little, Brown, Boston

Cato Institute (2002) The declaration of independence and the Constitution of the United States of America. Cato, Washington DC

Curtler H (2001) Recalling education. ISI, Wilmington

Harford T (2011) Adapt. Farrar, Straus, and Giroux, New York

Hayek F (1977) The use of knowledge in society. Institute for Humane Studies, Menlo Park

Hayek F (2009) A tiger by the tail. Ludwig von Mises Institute, Auburn

Hayek F (2011) The constitution of liberty. Chicago

Hazlitt H (1996) Economics in one lesson. Laissez Faire Books, San Francisco

Lane R (1984) The discovery of freedom. Laissez Faire, New York

McClanahan B (2012) The founding fathers guide to the constitution. Regnery, Washington DC

Menger C (1994) Principles of economics. Libertarian Press, Grove City

Mises L (1996) Human action, 4th rev ed. The Foundation for Economic Education, Irvington-on-Hudson

Mises L (2005) Liberalism. Liberty Fund, Indianapolis

Mises L (2011) Interventionism. Liberty Fund, Indianapolis

Newman J (1987) The idea of a university. Loyola, Chicago
O'Driscoll G, Rizzo M (1996) The economics of time and ignorance. Routledge, New York
Seldon A (2004) The collected works of Arthur Seldon, vol 7. Liberty Fund, Indianapolis

Chapter 2
The First Class: Subjective Value and Time Preference

Civil Engineers often claim to work for the public good. Their work between two environments, the physical and the economic, has been described in the following way: "[t]he function of engineering is to create utility in the economic environment by altering elements in the physical environment." (Thuesen and Fabrycky 2001)

In the physical environment, we know that this altering can only be achieved at a physical efficiency less than unity. However, for an activity in the economic environment to proceed successfully, the economic efficiency must exceed unity (i.e., must "make money"). Our question in this section is how to decide whether the economic efficiency of a proposed course of action is expected to exceed a numerical value of one.

The first class in my "Economy and Risk" course has been used to introduce students to the new concept of purposeful action in this economic environment. In later sections of this narrative, we will discuss a variety of decision models to achieve successful actions. For now, we must prepare an understanding of fundamental terminology and some basic tools to help make the decision whether to accept a proposed course of action.

2.1 The Meaning of Words

At this point, the instructor should instill in students the mandate that words have (or, should have) meaning. We will try to define our terms carefully and consistently. For example, at a very basic level, we must distinguish between "price" and "cost" as numerical descriptions of the value placed on some item. By "price" we intend to mean an exchange ratio. For example, the price of a pencil can be stated as, let us say, one-fourth of a "dollar." In an exchange ratio sense, we could equally correctly state that the price of a dollar is four pencils. The idea of a "pencil economy" will be developed in later classes by referring to the classic article "I, Pencil" by Read (1992).

© Springer International Publishing Switzerland 2015
R.A. Chadderton, *Purposeful Engineering Economics*,
DOI 10.1007/978-3-319-18848-5_2

The "cost" (or, opportunity cost) of an item is a different matter. For example, Thoreau (1965) wrote that "...the cost of a thing is the amount of what I will call life which is required to be exchanged for it, immediately, or in the long run." This means that only the actor can know the cost of anything. Cost is the most-valued thing foregone (Mises 1996). Thus, every student in a class may have faced the same price to attend college but no two of them may have faced the same cost; and, none of the students can know the cost to any other student. This is a fundamental situation that should be explained to each class.

Apparently, R.W. Emerson said that money often costs too much. This concept ("money") is often called a medium of exchange. At a later point, we will consider just what money is, what it is not, and how the "money function" evolved (Mises 1996; Menger 1994). For now, let it suffice to say that something called "money" can be used to determine whether an action is economically efficient and desirable. This will involve a process of "economic calculation."

2.2 Economic Calculation and Action

Economic calculation can be forward looking (prospective; ex ante) or backward looking (retrospective, ex post); but, as Mises demonstrated in the "economic calculation debate" of the early twentieth century, without "money" economic calculation would be impossible (Rothbard 2004). It would therefore be impossible to determine the economically efficient action to take and impossible to determine after the fact whether an action taken had been economically successful. For engineers, who are often involved with so-called public projects, the determination of economic success should be a crucial concern. We will argue that sound money is a prerequisite for this determination.

An "economic action" would tend to bring supply and demand into agreement. Without properly functioning money, there would be no prices and no possibility of economic calculation, hence, no "economics" at all. Supply and demand would not be in agreement. This means that both "suppliers" and "demanders" would be disappointed. The evidence of the Soviet Union could be invoked as proof.

This might also be an appropriate place in the class to introduce the difference between "limited resources" and "free goods." For example, oil was a nuisance-free good before it became a critical, limited natural resource. We should also consider resources as "limited" in two senses, physically and economically. This situation can be explained in terms of supply and demand.

2.3 An Economy Study Process

The following four "steps" in an engineering economy study are given by Thuesen and Fabrycky (2001), namely (1) creative, (2) definition, (3) conversion, and (4) decision. After the possible procedures and all of their respective inputs and outputs have been identified in terms of physical quantities, the "conversion step" calls for these to be put into a common denominator, "money" terms. We should also develop a project "time line" indicating when various events are expected to occur.

The example that I often use at this point in the class is construction of a dam/reservoir system. Some of the possibly conflicting capabilities used to justify the project might include water supply, hydropower, low flow augmentation, flood control, or recreation. The students are urged to look for "double counting" of so-called benefits when such competing uses are encountered. Also, the anticipated project "costs" (mobilization, clearing, construction) are noted to occur at different times during project development. Thus, the idea of time value is introduced as an important factor in the justification of the project.

Since the ultimate purpose of a proposed project is to satisfy future wants (an uncertain process), it will be imperative that we consider the time value of the money amounts. This time value will require the introduction of a so-called interest rate and an estimate of its numerical value.

So, what do we mean by "time value of money"? At the most fundamental level, money has time value because people prefer current goods over future goods. This means that current money should be more valuable than future money—we discount the future. The Austrian economists called this basic rate an originary rate of interest. Of course, an actual (market) interest rate would have to take into account additional factors such as risk and inflation that we will consider later in this narrative (Chap 4). Because the numerical value of the interest (discount) rate will influence the result of a decision rule, the selection of its value must be done with some deliberation. This effect will be demonstrated later (Chap. 8).

2.4 The Uncertainty of Predictions

We want to be clear that even though our numerical calculations from interest formulas seem precise, we cannot predict future conditions. We must contradict the positivist approach to economic analysis that attempts to mimic the predictions of mathematical physics. The application of the model of physics to economics has led economics astray because people do not behave like billiard balls (Rothbard 1979). Economic action is undertaken by purposeful individuals.

And so, we can play mathematical games (such as "sensitivity") by varying the numerical values in our cash flow equations. But, while we can visualize the changes in numerical results, we must not think that we are somehow predicting values in some absolute sense.

This attempt at prediction of the future is not a risk problem—it is uncertain as described by Mises (1996) because every situation in reality is unique; there can be no statistical summaries created from which probabilities for risk calculations can be derived. And, as developed by O'Driscoll and Rizzo (1996), information can be expected to evolve as a "project" proceeds; this is the effect of operating under a condition of uncertainty in "real time." And this is the degree of understanding that engineering students need to develop.

I often end the first class by asking the students to derive an interest rate by taking a poll. Ask: "How much would you want in return a year from now to lend another student $100 today?" There will be a range of "bids"; the instructor can summarize the results if the class is large enough. Pick the center value and demonstrate how that value might draw in some marginal borrowers and drive out some marginal lenders. More detail can come later, after the concept of operating at the margin has been explained.

References

Menger C (1994) Principles of economics. Libertarian Press, Grove City

Mises L (1996) Human action, 4th rev ed. The Foundation for Economic Education, Irvington-on-Hudson

O'Driscoll G, Rizzo M (1996) The economics of time and ignorance. Routledge, New York

Read L (1992) I, pencil. Imprimis 21(6):1–3

Rothbard M (1979) Individualism and the philosophy of the social sciences. Cato, Washington DC

Rothbard M (2004) Man, economy, state. Ludwig von Mises Institute, Auburn

Thoreau H (1965) Walden and other writings. Modern Library, New York

Thuesen G, Fabrycky W (2001) Engineering economy, 9th edn. Prentice Hall, Upper Saddle River

Chapter 3
A Few Basic Concepts: Setting the Stage

The purpose of this book is to present supplemental material that I have incorporated into my course called "Economy and Risk" built around an engineering economics textbook. These are materials that I believe every engineering graduate should appreciate.

The objective of my course is understanding and analysis, not speed. Therefore, financial calculators and spreadsheets for numerical calculations are not emphasized, although students tend to use these physical tools anyway. Emphasis is placed on the diagrams, formulas, and constants used to solve a variety of numerical cash flow problems. Students should also learn to appreciate the non-mathematical aspects of economics from in-class discussions.

3.1 The Most Important Equation

There is only one equation necessary for understanding the most important, basic concepts. This is the single payment compound amount or future worth formula, Eq. 3.1.

$$F = P(1.0 + i)^n \tag{3.1}$$

Reference to Eq. 3.1 will be made at many points throughout this narrative. It can be applied in several different situations to accomplish several different objectives. Students should remember Eq. 3.1 and "understand" it, to exhaustion!

3.2 The Components of the Course

I describe the composition of my course as being made up from four "threads" that run throughout the course.

© Springer International Publishing Switzerland 2015
R.A. Chadderton, *Purposeful Engineering Economics*,
DOI 10.1007/978-3-319-18848-5_3

1. Economics: This is a non-mathematical branch of the social sciences; the main sources of information are Mises (1996), Rothbard (2004), Sowell (2000), and Hayek (1977).
2. Engineering Economy: This is fundamentally cash flow analysis following a textbook such as *Engineering Economy* by Thuesen and Fabrycky (2001).
3. Risk and Uncertainty: Risk is described as an application of probability concepts. However, emphasis must be placed on the fundamental uncertainty of all economic predictions. References include Wildavsky (1988), O'Driscoll and Rizzo (1996), Hayek (1991), and Mises (1996).
4. Current "Issues" and Historic "precedents": Topics are drawn from political economy and the media, both current events (reporters, journalists) and related commentary. Recommended reference materials include Sowell (2000), Williams (2008), and Hazlitt (1996).

It is virtually impossible to "weave" a smooth fabric from these four, disparate subjects. The benefit from trying is the light they shine collectively, producing a more complete picture of the purposes, analyses, and results of calculations known commonly as engineering economy.

3.3 Perspective of the Course

At this point in our discussion, students are encouraged, actually expected, to "question the unquestionable." Examples of subjects that are presumably beyond questioning, because they are universally understood, include inflation, interest, money, taxes, price controls, minimum wages, and even rent control. The point is that so much "common knowledge" about the aforementioned "unquestionable" subjects that students have learned from the media or politicians is fundamentally inaccurate, yet continues to be repeated by politicians and reporters. During any given semester that the course is offered, there occur numerous incidents related to these topics that can be drawn upon for in-class discussions. All that the instructor must do is to be alert to misleading pronouncements prominent in the media.

It must be made perfectly clear that none of this questioning is political lobbying, except in a most general way. I have been accused of being a Republican and of being a Democrat but neither accusation is correct. There is no "party line" being promoted. However, my courses do discuss reference materials and events from an unusual perspective, namely the classical liberal position. One excellent exposition of this view has been presented by Mises (2005).

Students are explicitly told that they do not have to agree with presented arguments, but they do need to understand what the arguments are. I always document the sources of my reference materials (truth in advertising, again) and handouts of crucial articles to be studied. Thus, students will know how arguments have been developed. One of the best references is Hazlitt's *Economics in One Lesson* (1996) that I have assigned for summer reading prior to my class for several years.

3.4 The Crucial Lesson

As Hazlitt (1996) explains the situation, *"The art of economics consists in looking not merely at the immediate effects of any act or policy; it consists in tracing the consequences of that policy not merely for one group but for all groups."* In other words, the long-term consequences for everyone should take precedence over the short-term benefits for special interest groups.

The *Lesson*, more than 68 years since its publication, has not been learned. The book could be rewritten even today, because the same errors it explains continue to be repeated. For example, legislation to raise the minimum wage was on the ballot in several states during the 2014 election cycle. At least the few graduates of my class have some awareness of the questions raised by the *Lesson*, one of which is the error of minimum wage laws.

3.5 The Coordination Problem

I take a #2 wood pencil into the first or second class meeting and ask the students, "What does it take to make a #2 pencil?" What goes into it? Compile a list of all their answers which will probably include:

- materials (wood, metal, graphite, rubber, paint);
- labor ("how much"; "what kind" are always additional questions);
- machinery (from where?, capital; what kind again);
- transportation (method?); and
- distribution (how?)

In some cases, they might even add "thought" and "time." But if they don't, the instructor should add them to the list. We want to demonstrate how complex the processes are to supply as simple an object as a pencil.

The conclusion demonstrated by this exercise is the truth of Leonard Read's article "I, Pencil" (1992) that "No one knows how to make a pencil." So, even the simplest of consumer items demonstrates the difficulty of coordination of all of the individual actions required to create and supply that item. Implied in this story is the importance of price signals to accomplish the required coordination of activities of many individuals who know nothing of the multitude of others involved in making the pencil. The crucial significance of these signals justifies a more complete explanation of what will be called "information transfer" later in the course (Chap. 7).

I distribute copies of "I, Pencil" with permission from *Imprimis* so the students can acquire a more complete understanding of the coordination problem involved. I also mention that Thoreau's family business was pencil making.

References

Hazlitt H (1996) Economics in one lesson. Laissez Faire Books, San Francisco
Hayek F (1977) The use of knowledge in society. Institute for Humane Studies, Menlo Park
Hayek F (1991) The fatal conceit. Chicago
Mises L (1996) Human action, 4th rev ed. The Foundation for Economic Education, Irvington-on-Hudson
Mises L (2005) Liberalism. Liberty Fund, Indianapolis
O'Driscoll G, Rizzo M (1996) The economics of time and ignorance. Routledge, New York
Read L (1992) I, Pencil. Imprimis 21(6):1–3
Rothbard M (2004) Man, economy, state. Ludwig von Mises Institute, Auburn
Sowell T (2000) Basic economics. Basic Books, New York
Thuesen G, Fabrycky W (2001) Engineering economy, 9th edn. Prentice Hall, Upper Saddle River
Wildavsky A (1988) Searching for safety. Transaction Publishers, New Brunswick
Williams W (2008) Liberty versus the tyranny of socialism. Hoover, Stanford

Chapter 4
Price and Cost

One of the most important points of misunderstanding that should be corrected is the confusion between the terms "price" and "cost." These terms are often associated with the concepts of "utility" and "value." The term "utility" might be described as an estimate of the ability of a good to satisfy human wants. The related term "value" could be expressed as "an appraisal of utility in terms of a medium of exchange" (Thuesen and Fabrycky 2001). Both utility and value are attributed to an object by an individual; they are not inherent in the object.

4.1 Human Action

The individual essentially assigns a value/utility to an object. However, the only way for such value/utility to be meaningful, or actually to develop or be exhibited, is if that individual has the ability to take action. At this point in the class, we must explain what we mean by "action" in this context.

The Austrian view of human action is the effort made to remove felt uneasiness; which becomes the starting point for the development of economics, or catallactics, as presented by Mises (1996). For an individual to remove this felt uneasiness implies the ability to carry out the required action; then, the "value" (utility) might be realized.

© Springer International Publishing Switzerland 2015
R.A. Chadderton, *Purposeful Engineering Economics*,
DOI 10.1007/978-3-319-18848-5_4

4.2 Value Theory

In this course, we consider the following three theories of value:

(1) Objective Value: The object somehow contains inherent value. The object is "valuable" even without anyone "valuing" it. But, if no one values it, would it be an element of economic action; one could reasonably conclude "no."

(2) Labor Theory of Value: Simplistically stated, only labor can impart value to an object. Also, simplistically stated, the consumer would be willing to pay more (implying a greater value) for an identical object made "laboriously" by hand as compared to one made by machine (say, a wood pencil). It seems that this "theory" attributed to Marx can be easily discredited by having students vote in class.

(3) Subjective Value Theory: Each individual assigns his own estimate of value and creates an ordering of relative values, an ordinal ranking of preferences. Then, the various interacting individuals bid by exchanging items. The multitude of exchanges by many individuals participating in a world having a medium of exchange essentially creates a "price structure."

It seems that only the subjective value theory supports the Austrian concept of "human action" without obvious contradictions.

4.3 Information Transfer

In a world of social organization by the division of labor, each individual participates both as producer and as consumer over time. To explain this situation more clearly, the two "functions" are often considered separately. Austrian economists generally explain the process that I label "information transfer" as two related processes.

The flow of "goods" is from higher order producers' goods to successively lower orders of producers' goods until, finally, a consumers' good (first order) is created. The more "roundabout" or time-consuming a particular production sequence is, the more (higher) levels of producers' goods would be involved. We can elaborate on this production process by referring to the pencil case (Chap. 3) and filling in more details about the many coordination processes required.

Note, however, that "information" passes up the production process from the ultimate consumer to the successively higher levels of production. Thus, we argue that it is the prices that consumers are willing to pay for consumer goods that will ultimately determine the allowable "costs of production," including labor, that can be sustained and not vice versa. We hear too often from the media that "rising costs of production are pushing up consumer prices." This is the kind of error that should be contradicted in this class. What is actually happening is that consumers willing to pay higher prices allow producers to incur higher costs. The instructor can refer to Rothbard (2004) for a more detailed explanation of the production and information exchange processes.

Now, since voluntary exchanges among the many participating individuals are the determinants of "prices," we need to consider what "elements of exchange" operate to determine which exchanges will actually occur. We remember that we are talking about indirect exchanges occurring in an economy using money as the medium of exchange. We will now investigate just how this money came into being and whether it is legitimate money.

4.4 On the Origin of Money

The purpose of this section is to give a brief summary of the Austrian theory of the origin of money. This is the explanation that I have used in my classes and should be an adequate introduction for engineering students. They can be referred to the complete presentations of Menger (1994) or Mises (1981, 1996).

Mises basically discounted theories of the origin of money by government decree or by explicit compact among citizens. Menger's theory of the origin of money was irrefutable (Mises 1996).

Acting individuals have everywhere found indirect exchange to be more convenient than direct exchange in their pursuit of their own self-interests. They have discovered the most useful mediums for exchange, usually cattle at first (Menger 1994). As the number and complexity of economic actions increased, more convenient commodities were found to facilitate exchanges. Gradually, market processes led to the selection of the most marketable commodities to serve this purpose; these became universally used as the medium of exchange, or money. The resulting monetary function increased the value of the commodity and made it even more desirable as a medium of exchange.

Since the use of the most marketable commodities in exchange was in the economic interest of every acting individual, there was no need for legislative compulsion or citizen compact to establish these as money. Menger (1994) concluded that money was neither an invention of the state nor the product of a legislative act. Thus, money is a prime example of an evolved social institution explained by Hayek (2011).

Ultimately, pieces of precious metals such as gold or silver came to be accepted as the most useful items to facilitate transactions. However, the difficulties of guaranteeing weight and fineness of these pieces of metal were problems in each exchange.

Governments eventually assumed the tasks of stamping and milling metallic coins to guarantee weight and fineness (value) of pieces of precious metal of various sizes. Names for coins typically referred to their weight, for example, pound in England or livre in France (Menger 1994). Of course, the temptation for government officials to debase the coinage then became a new problem for the citizens that will be considered again in Chap. 6.

I have reminded students that Sir Isaac Newton was "Master of the Mint" at London for many years. His first task was to recall old, clipped, and thus overvalued coins that were causing unrest among the population and to replace them with new, milled silver coins of true weight (Andrade 1958). This historical event could provide an excellent case study in the operation of Gresham's Law as described in Greaves (1974).

4.5 Preference and Action

We must also remember that each individual has created an ordinal "ranking of preferences"—an ordering of choices almost like a stack of cards—in terms of subjective values, not "prices." No one can know any other individual's order of preferences. It is not clear whether a person could write down his own list in total or is even consciously aware of what the list would be because it would be in a constant state of flux as new choices became available (O'Driscoll and Rizzo 1996). The clear political implication is that no one can be in the position of knowing what everyone needs, or prefers, or how much anyone would be willing to pay.

I often use the example of college tuition to contrast "price" and "cost" and to explain the planner's impossible contradiction. Presumably, every student in the class has incurred the same tuition price by some combination of financing methods. But, because of many different rankings of individual preferences, no two would have accepted the same "opportunity cost." Every individual in the class would experience a different tuition cost. No student would be able to determine the cost that any other student is willing to pay. Clearly, a "planner" could never know the cost preferences for all of the students! We might notice that this conclusion would be particularly relevant to young, healthy students facing the new "healthcare" mandate. Many of them might reasonably want to use their healthcare dollars for something of more immediate concern.

While it is often maintained that items exchanged are of equal value, this is not true. Unless both parties to a voluntary exchange have "removed felt uneasiness"—that is, have acquired something of greater value than that given up—there would be no reason for the exchange to occur. If one adds transaction costs, this conclusion would be even more obvious. This is the concept of mutual benefit that is essential for voluntary exchanges to occur. Each party to an exchange must feel better off after the exchange has occurred.

For an individual actor, the second item of the same kind will be less valued than the first. We must be very careful that these two items are exactly identical! For example, a bushel of corn in Iowa is not the same as that bushel of corn relocated to New York City and would not be valued equally by a person in Iowa and a person in New York City. This line of thought can lead to the "law of diminishing returns" or "diminishing marginal utility." That is to say, for example, that a student's first calculator might be very valuable to that student, but having a second calculator would be less valuable than the first. In fact, the second calculator might be several items below the first calculator on the student's ordered list of preferences. I give some ideas for preferences that might intervene between the two calculators, including "leisure," because students should understand that leisure can become an opportunity cost.

4.6 Some Categories of Cost

The calculator example can be extended to demonstrate the idea of <u>sunk costs</u>. Imagine that you (student X) just bought a new calculator for, say, $200 at the University bookstore. The next day, or soon after, student X learns about an even better calculator that "costs" only $100, but does much more than the $200 one. One question that you must answer is "How much is my 'old' calculator worth?"

I ask the class to bid to buy the "old" calculator that student X bought. The highest bid will probably be less than $100, in this case. Poor student X has suffered a great loss, aren't the other students taking advantage of the situation? Hardly, they are actually offering a price more than zero. Student X has incurred a "sunk cost," due to no fault of his or her own. It is psychologically difficult to ignore it. But, human action must be directed to the future, not the past. This example also demonstrates the idea that all economic data (prices, in this case) are historical values—based in the immediate past. Future, expected prices are "guesses" based on the immediate past (Mises 1996). The difficulty of uncertainty arises because engineering economy seeks to predict the future economic effects of current decisions, but economic data are historical values, past data, only.

We are working in an environment of indirect exchange using money as the medium of exchange. Prices and costs are stated in terms of the money unit; I sometimes use the "University Unit" to emphasize that these numbers are really just exchange values.

Some additional cost classifications that can be discussed at this time include first cost, operation and maintenance, fixed cost, variable cost, and marginal cost. These costs can be shown on a simple sketch, leaving off number scales to emphasize the uncertain nature of these estimates.

Important observations about the shape of this sketch could include economy of scale, decreasing average cost, and decreasing marginal cost. At this point, I want us to emphasize the fact that "We always operate at the margin," meaning that all of our decisions are marginal decisions. One pertinent example the instructor can give would be to compare an individual's average income tax rate (maybe 10–15 %) with the marginal tax rate (probably 25–28 %). This should make a lasting impression, even on undergraduate students!

The sketch would also allow us to ask the classical question: "Why do diamonds cost more than bread, when bread is obviously so much more essential?" The classical economists apparently labored over this question. The answer is that the total value of all the bread in the world is more than the total value of all the diamonds. But, no one would be in the position to choose between all of the bread and all of the diamonds. There is much more bread, but the marginal value of a pound of diamonds is greater than the marginal value of a pound of bread.

We previously considered the "sunk cost" of a purchase in the calculator case. I also use the example of a house that had lost a portion of the purchase price on the market. The terms of the existing mortgage, left unchanged, cause a sunk cost to the buyer. We will learn how to calculate repayment of this loan later. For now, this current event should be of interest to these future home buyers.

4.7 Interest and Interest Rates

Engineering economy textbooks describe interest in different ways. Some say that interest is rent paid for the use of the money (Blank and Tarquin 2008) or the cost of having money available for use (Park 2013). Other descriptions of interest include the rate of growth of capital (Thuesen and Fabrycky 2001), the "earning power" or "purchasing power" of money (Park 2013), and the rate at which money increases in value (Au and Au 1992). Causes of interest can include risk of loss, administrative expenses, gain after accounting for inflation, and foregone opportunity of investing elsewhere.

While some of these explanations imply it, none is explicit about the "originary rate of interest" which derives from people's pure time preference for current goods as opposed to future goods.

The mathematics of interest rate formulas does not depend on the causes of interest, but only on the numerical value used in Eq. 3.1. However, for a deeper understanding of "time value," students need to know the constituent parts of the interest rate.

Mises (1996) demonstrated that time preference is an inherent (praxeological) category of human action simply because people are aware of time. They will always value present goods more than future goods because they are here now; future goods are not. If this were not true, a person would never consume in the present, but would always wait for some future that would never be the "present."

Thus, the originary interest is "the difference between the present values of present and future goods." In addition to this originary interest, the market rate of interest must include an amount for risk (uncertainty of repayment), inflation (changing value of the monetary unit), a price premium (changing value of goods), and possibly an amount for profit.

The interest rate is then used to calculate present values of goods from anticipated future values (or, vice versa) by using Eq. 3.1. And yet, we still need to investigate what we mean by a monetary unit, money. Is the monetary unit a creation of government or a mere convention? This question was addressed by Menger's (1994) theory of money in Sect. 4.4.

In terms of supply and demand, what would happen if the charging of interest were to be outlawed? This is a case of making "usury" criminal, as during the Middle Ages. Stories exist of people risking their lives to borrow money with interest during those times. As can be shown in a simple supply and demand diagram, for a 0 % price for interest, supply would be zero and demand would be essentially unlimited.

4.8 An Introductory Investment Decision

This section uses a hypothetical example of an investment decision to elaborate basic concepts. First, we must emphasize that this example takes place in an untaxed, non-inflationary world. This is the typical case for the early chapters of most engineering economics books. Then, we can describe the facts of the case.

The owner borrowed a significant amount of money at a specified rate of annual interest to buy capital equipment for his small business. The company experienced increased net income while reducing the unit price of its product because of the machine. Textbook authors often describe this as an illustration of the earning power of money. The owner has more net income, and customers get lower prices.

The owner apparently based this decision to borrow money on a prospective calculation. The success of that decision was apparently shown by a retrospective calculation. But, again, there was no inflation or taxation involved. The case suggests a need for several elaborations.

We could begin by mentioning the Luddite claim that the acquisition of capital equipment caused unemployment of manual laborers and their destruction of the machines they blamed for causing unemployment in particular industries. The question for our students might be whether the investment necessarily caused unemployment. We can argue that the customers' savings in this case could be expected to create employment of a different kind; they have the capacity for additional expenditures. A lower supply price curve would indicate additional demand.

This example can be extended by asking "What made the owner's loan possible?" The idea is to introduce the critical role of previous savings for investment to occur. If the bank did not have deposits from other customers, it could not make the loan. This conclusion leads to the further question of effects of fractional reserve banking.

We can also use this hypothetical case to ask how borrowing from a bank would work. I tell students that two pieces of paper exchange hands. "You give a signed note, a mortgage, and get a check, a claim for dollars. Where did the dollars come from?" Another question is to ask what happened to the idea of "mutual benefit" of this transaction?

There must be a difference of opinion about the terms of the loan between the borrower and the lender because if there was indifference on their parts, the transaction would not occur. Apparently, the lender must think that the correct rate of interest is less than the loan rate and the borrower must think that the correct rate is more than the loan rate. This difference of opinion allows both parties to expect increased value from the exchange, even after accounting for the transaction costs.

4.9 A Brief Introduction to Banking, Money, and Credit

It might be a good idea to ask about the effect of fractional reserve banking on prices. I start by describing the progression from the warehousing of commodity money to the use of paper money. In short, banks evolved from warehouses. Merchants stored commodity money in warehouses to avoid the dangerous transportation of heavy commodity money for trading purposes. Deposit receipts from trusted warehouse operators developed monetary properties. The warehouses evolved into banks and the receipts evolved into demand deposits that could be redeemed on demand by the bearer of the receipts for commodity on deposit.

If a warehouse owner created receipts for more than the deposits of commodities in his warehouse, there was essentially fractional reserve banking. The next question for students is what would happen if all the holders of demand deposit notes tried to claim commodities? It should be easy to see that only those first in line would be able to redeem the deposit claims. But, in a world of paper money and paper receipts, who would get the "real" dollars from a fractional reserve bank? These questions can be viewed as an extension of the Austrian theory of money in Sect. 4.4. I mention the bank run from the movie, *It's a Wonderful Life*, that most students will have seen. Mises' classic book (1981) gave a thorough explanation of these questions.

This is also an appropriate time in the class to consider the purchasing power of money. The structure of prices is affected, not only by productivity, but also by price controls (supports and ceilings) and by inflation. While the current usage of the term "inflation" means price increases, the correct definition is an increase in the supply of money. So, we could conclude that fractional reserve banking would promote price increases and further distortion of the structure of prices. The situation would increase the likelihood of making poor decisions based on honest prospective calculations.

We also note that the increased net income for the owner in the previous example would be substantially reduced by taxation. And, worse yet, the value of the remaining income would be reduced, possibly significantly, by inflation. The effects of taxes and inflation will be added as students progress through the course.

To conclude the hypothetical investment example, I introduce what I call "Rich Uncle Economics 101." The first lesson comes from the question, "If the company owner had the cash (from his rich uncle), would he be wise not to borrow the money?" I tell the students to keep their respective answers to themselves for now and to have reasons for either answer.

References

Andrade E (1958) Sir Isaac Newton: his life and work. Doubleday & Company Inc, Garden City

Au T, Au T (1992) Engineering economics for capital investment analysis, 2nd edn. Prentice Hall, Englewood Cliffs

Blank L, Tarquin A (2008) Basics of engineering economy. McGraw-Hill, New York

Greaves P (1974) Mises made easier. Free Market Books, Dobbs Ferry

Hayek F (2011) The constitution of liberty. Chicago

Menger C (1994) Principles of economics. Libertarian Press, Grove City

Mises L (1981) The theory of money and credit. Liberty Fund, Indianapolis

Mises L (1996) Human action, 4th revised edn. The Foundation for Economic Education, Irvington-on-Hudson

O'Driscoll G, Rizzo M (1996) The economics of time and ignorance. Routledge, New York

Park C (2013) Fundamentals of engineering economics, 3rd edn. Pearson, Upper Saddle River

Rothbard M (2004) Man, economy, state. Ludwig von Mises Institute, Auburn

Thuesen G, Fabrycky W (2001) Engineering economy, 9th edn. Prentice Hall, Upper Saddle River

Chapter 5
Equivalence

This discussion of "equivalence" assumes that the students have previously learned the basic interest formulas and diagrams for cash flow analyses. The engineering economy textbook used for the course will include the complete formulas and tables of interest factors using a version of the familiar notation shown below.

Following is a listing of shorthand versions of the fundamental interest formulas that are emphasized in this class. All are represented for cash flows of discrete payments with discrete compounding of interest. For these formulas, P is a present amount, F is a future amount, A is an equal annual amount, and G is an annual incremental amount.

1. Single payment compound amount factor (F/P).

$$F = P * (F/P, i, n)$$

2. Single payment present worth (PW) factor (P/F).

$$P = F * (P/F, i, n)$$

3. Equal payment series compound amount factor (F/A).

$$F = A * (F/A, i, n)$$

4. Equal payment series sinking fund factor (A/F).

$$A = F * (A/F, i, n)$$

5. Equal payment series capital recovery factor (A/P).

$$A = P * (F/P, i, n) * (A/F, i, n)$$
$$A = P * (A/P, i, n)$$

© Springer International Publishing Switzerland 2015
R.A. Chadderton, *Purposeful Engineering Economics*,
DOI 10.1007/978-3-319-18848-5_5

6. Equal payment series PW factor (*P/A*).

$$P = A * (P/A, i, n)$$

7. Uniform gradient series factor (*A/G*).

$$A = G * (A/G, i, n)$$

In all cases, the interest rate, i, must be the appropriate rate for the time step involved; the number of these time steps is n. There are other formulas available, such as those including more frequent compounding of interest. Such cases can be handled by using an effective interest rate in the basic formulas, above. We also note that with appropriate trial-and-error procedures, Eq. 3.1 could be used to solve any of the cases listed above. The reason that two equations are shown for the term (*A/P*) is to demonstrate that the notation used for the interest factors allows for algebraic cancellation, that is $(F/P) * (A/F) = (A/P)$.

In this section, we will investigate methods to compare alternatives. This will require the use of a common basis for comparison which in economy studies will be money. The elements of concern will be amounts of money, the time when these amounts occur, and the rate of interest (discount). Obviously, these three elements can be visualized by an appropriate "cash flow diagram," and they can be related and manipulated by the various interest formulas or interest "factors" studied previously.

5.1 A Simple Decision Problem

I begin the in-class presentation of "equivalence" by offering a simple choice. The students are told, or asked, to choose either (a) $12,500 "now" or (b) $2000 per year for each of ten years, starting one year from "now." The basic question comes from the Thuesen and Fabrycky (2001) textbook, but here it is elaborated as an example of "Rich Uncle Economics." Imagine that your uncle has given you, each individual student, the choice of either (a) or (b). But, once you choose, there is no changing your mind!

I ask for a show of hands for each choice and put the results on the board, e.g., (a) n and (b) m. Then, I ask how the students have made their choices; there are usually a variety of reasons given. Then, I ask what "decision rule" the students had used. Sometimes, a student will suggest a rule such as PW, but usually not. They just picked. Someone will ask what the interest rate is. I reply that I don't know, what do you think? It's your choice to make. The instructor can point out that this is the choice you make when you win the lottery. But the question of how to choose remains.

The point is that each individual needs a decision rule whether they are conscious of it or not. This question of "how to decide" can be used as an introduction to the meta-decision problem of deciding how to decide. In this first case, I choose PW for the class. The "equivalence" equation is now:

$$\text{"equivalent"}P' = \$2000(P/A, i, 10)$$

or the "discounted" PW of choice (b) is P', which clearly will depend on what discount rate the individual student picks. But, if $P' < \$12{,}500$, you would choose (a), and if $P' > \$12{,}500$, you would pick (b). The following table that includes some example rates of interest is shown after the students make their choices.

$i\,(\%)$	$P/A, i, 10$	P'	Choose
12	5.6502	$11,300	(a)
11	5.8892	$11,778	(a)
10	6.1446	$12,289	(a)
9	6.4177	$12,835	(b)
8	6.7101	$13,420	(b)
0	10.00	$20,000	

There seems to be an interest/discount rate at which options (a) and (b) are equivalent, meaning that they both reduce to the same PW. We can prove that the "break-even" interest rate is 9.614 % annual by a rate of return calculation with $P' = \$12{,}500$ and $(P/A, i, 10) = 6.2500$. However, if someone chooses option (b), the apparent (calculated) P' would not actually occur. The actual payments would be $2000 (I call them "counted out" dollars) at the end of each of ten consecutive years. This observation clearly shows the difference between the <u>actual receipts</u> and the <u>equivalent amount,</u> P'. We also note that for $i = 0\,\%$, $P' = \$20{,}000$, the upper limit which is the number of actual dollars that choice (b) would bring to you.

We note that each person might decide differently or, in this case, divide between the two choices. Also, the actual "worth" of your choice might prove not to be what you predicted. For example, the discount rate you used might be wrong or it might have changed during the ten years. Your retrospective calculation does not confirm your prospective calculation!

We set "equivalent" $P' = \$12{,}500$ in our equation and solved for $i = 9.614\,\%$ (annual). We could say that option (b) is "returning" 9.614 % annual interest. Since we chose option (b), option (a) is a "foregone opportunity" and we can think of it as an investment that earns 9.614 % annually. So, if the market rate of interest is more than 9.614 %, we should choose option (a) and invest $12,500 received now at the market rate. We can also ask whether the students think that taxation would affect their choices. Maybe it should, and we will consider that question later (Chap. 9).

5.2 An Example Application: Creating a Retirement Account

We could say that most, if not all, cash flow calculations are equivalence cal-culations. One very important application for students would be to estimate the required deposits to create a retirement account. Therefore, we develop the follow-ing example in class.

Start with a target amount at some future time and an assumed interest rate. For ease of calculation, work with a future amount of $1000 (units) and an assumed interest rate, let us say 8 % annually. Then, we can calculate the required annual deposit to accumulate $1000 after 5 years:

$$A = \$1000(A/F, 8\,\%, 5) = \$1000(0.1705)$$
$$A = \$170.50(\text{at the end of each of five consecutive years})$$

$$F = \$170.50(F/A, 8\,\%, 5) = \$170.5(5.867) = \$1000.32$$

Repeated use of Eq. 3.1 will confirm the F/A calculation.

We can ask what amount today (P at time 0) would be equivalent to $1000 five years later at 8 % annually. Thus,

$$Po = F/(1+i)^n = \$1000/(1.08)^5 = \$680.58$$

We now have three equivalent cash flows; namely, Po, F, and A, all evaluated for 8 % annual interest.

Next, we can ask what annual amount can be withdrawn from the accumulated F for each of five years if 8 % interest continues to be valid? This is a capital recovery calculation using $(A/P, 8\ \%, 5) = 0.2505$ so that

$$A = \$1000(0.2505) = \$250.50 \text{ annually for five years.}$$

We can ask the students to put the two cash flows onto a single timeline to show our retirement plan as a cash flow diagram complete with calendar dates over the ten-year period.

Note, this example can be amplified in several ways. Have the students calcu-late the number of dollars deposited ($852.50) and the number of dollars with-drawn ($1252.50) but remind them that this comparison is between numbers of "dollars" and not "values" because of the passage of time.

Students should be able to calculate the account balance after each transaction and tabulate these amounts. Ask them to prove that the first A series (deposits) does accumulate to $1000 using only Eq. 3.1. Similarly, ask them to prove that the second A series (withdrawals) just depletes the account after five equal withdraw-als, again using only Eq. 3.1. These calculations are shown below.

Year #	Balance before transaction	Transaction	Balance after transaction
0	0	0	0
1 ("today")	0	$170.50	$170.50
2	$184.14	$170.50	$354.64
3	$383.01	$170.50	$553.51
4	$597.79	$170.50	$768.29
5	$829.76	$170.50	$1000.26
6	$1080.28	−$250.50	$829.78
7	$896.16	−$250.50	$645.66
8	$697.31	−$250.50	$446.81
9	$482.56	−$250.50	$232.06
10	$250.61	−$250.50	$0.11

We can emphasize that a similar set of calculations could be used to amortize any loan, such as a home mortgage. I include such a mortgage loan amortization calculation in class.

5.3 Equivalence with Working Capital

Strictly speaking, working capital could include inventory, accounts receivable (outstanding anticipated income), and cash on hand The intended lesson is for students to learn that if "working capital" is ignored, the cost of a project will be underestimated. Let us say that you expect to have a need for some amount, call it Pw, as cash on hand during a project of several years duration. You also expect to recover all of Pw at the end of the project.

I view this situation as a case of "Grandfather Economics" referring to the Great Depression years when "grandfathers" put cash money under the mattress so that it would not be lost in a bank failure, for example. Textbooks typically show the solution of this problem as an equivalent cash flow: Pw as a "cost" at time "0" and Pw as an "income" at time "t". The "equivalent annual cost" is then given by:

$$A = -Pw(A/P, i, t) + Pw(A/F, i, t) = -i * Pw$$

The negative sign indicates a net equivalent, not real, cost that could be calculated and included with other project costs. Doing so would make the project look less attractive and should be included in the decision process.

I have the students prove algebraically that

$$(A/P, i, t) - (A/F, i, t) = i$$

This result is also evident from scanning the "interest tables" in any textbook and is true for any i or t. Thus, there is a "foregone opportunity" or uniform annual cost (UAC), equivalent to the interest that Pw could earn, when "grandfather" stores cash money in his mattress!

Another way that I have students look at working capital is from the lender's view and involves a "deposit" or "loan" of Pw made at time (0). After one time period, say a year in this case, the accumulated amount due to the lender (a "draw" or "payment" due) would be equal to $Fw = (1.0 + i) * Pw$. Therefore, after one time period the, "opportunity cost" of not earning interest would be:

$$"A" = Fw - Pw = [(1.0 + i) - 1.0] * Pw = i * Pw$$

And, this annual "opportunity cost" of the capital would be the same for each year of the project duration. Thus, "grandfather" is giving up the interest that Pw could earn by storing the actual cash money. Presumably, the peace of mind of having the cash available at all times is worth more than the foregone opportunity of interest that could be earned. Possibly, a higher rate of interest would draw some of the hidden cash from under those mattresses.

5.4 Summary of Principles of Equivalence

I end this segment of the course with a few reminders about some principles of equivalence:

1. The actual worth of what you choose may not be what you predicted (prospective versus retrospective calculations).
2. Remember the time conventions for interest formulas.
3. Only money amounts adjusted to the same time can be added.
4. The reference time can be any convenient point (we used the present).
5. Equivalent cash flows will be equivalent at any reference time.
6. The problem could be posed in different ways, e.g.,

   (a) Given P and i, what F would be equivalent after n years?
   (b) Given P and F, n years later, what i was actually earned?

Finally, remember to distinguish clearly between numerical predictions and actual money amounts. This final statement cannot be repeated too often!

Reference

Thuesen G, Fabrycky W (2001) Engineering economy, 9th edn. Prentice Hall, Upper Saddle River

Chapter 6
Inflation

Why should we discuss inflation with engineers? Hasn't inflation been kept "under control" in recent years because of actions by the Federal Reserve? These are clearly legitimate questions that deserve answers.

On a practical level, students must be able to account for inflation in cash flow analyses to arrive at correct numerical results. More importantly, they should have a basis in theory to understand how and why inflation happens.

Most of us have been led to believe that inflation is caused by external forces that must be fought by the government. We have also been told that a little inflation is good for us and that the national (federal) debt is no worry because we owe it to ourselves. Therefore, another purpose of this section is to contradict these errors of our common knowledge.

6.1 Austrian Explanation

Hayek explained the situation in the following way:

> Those who wish to preserve freedom should recognize that inflation is probably the most important single factor in that vicious circle wherein one kind of government action makes more and more government control necessary. For this reason all those who wish to stop the drift toward increasing government control should concentrate their efforts on monetary policy. There is perhaps nothing more disheartening than the fact that there are still so many intelligent and informed people who in most other respects will defend freedom and yet are induced by the immediate benefits of an expansionist policy to support what, in the long run, must destroy the foundations of a free society. (Hayek 2009).

A consequence of the policy of expansionism is fear of hyperinflation, a phenomenon that Mises (1981, 1996) called a "crackup boom." Very few citizens really understand inflation at a fundamental level. Inflation is more complicated than simply increasing prices. Our purpose should be to distinguish clearly

© Springer International Publishing Switzerland 2015
R.A. Chadderton, *Purposeful Engineering Economics*,
DOI 10.1007/978-3-319-18848-5_6

between the phenomenon of increasing prices and the process of inflation. This is a case where we could have a beneficial effect on public policy by giving a fundamental understanding to our students.

Inflation is a policy. Hayek (2009) called it a "tiger by the tail." Friedman (1994) wrote that inflation "is always and everywhere a monetary phenomenon." Even Keynes said that it is "a colossal muddle, having blundered in the control of a delicate machine, the working of which we do not understand". (quoted by Harford 2011). Mises presented detailed theoretical explanations (1981).

Inflation fundamentally eliminates debt. Who is the greatest debtor? Who controls inflation? If the answers to these two questions are the same, there is a danger to the economy.

The accumulation of capital supports improving living standards. Inflation distorts the structure of prices, thus confusing (falsifying) price signals, leading to capital consumption. Taxation of inflated values aggravates the phenomenon of capital consumption. These effects will be considered further in Chap. 7.

6.2 Introductory Examples

Ultimately, inflation restricts attempts to improve living conditions or, in extreme cases, even causes living conditions to deteriorate by reducing purchasing power and causes distortion of exchange ratios between money and goods. I introduce these effects in class using the following personal experiences.

Example 6.1 **A Personal Introduction to Effects of Inflation; Index Numbers**.

Christmas 1964: My parents bought a Post Versalog slide rule from our college bookstore for $28.50 for my engineering studies. A typical textbook cost about $20, or less, at the time.

1972: The HP 35 pocket calculator was priced at $395. The construction company where I was working bought one.

1974: As a graduate student, I bought a TI 20 pocket calculator for $169.95, by mail order.

"Today": Calculators priced around $15 can do much more than the HP 35 or TI 20. The latest price for my slide rule on the Web was $395, if available. Textbooks made of cheaper materials now have prices in the hundreds of dollars.

If we calculate price ratios for the slide rule and "equivalent" calculators over the time period (namely 1964 until about 2012), we see that

"calculator ratio" ~1/20
"slide rule ratio" ~14/1

(And, of course, the "book ratio" would be more than ten.)

The observation, or lesson, intended here is that prices do not move together the way observations of some index number (such as the Consumer Price Index

(CPI)) over time would indicate. Prices of some items decline in the face of rising CPI type numbers. What should students take away from these observations?

Using the common index number, (CPI), for the same time period as the previous example, we see that

CPI(1964) ~94
CPI(2011) ~676

for a CPI ratio of about 7.2.

The ideas that I want to share with students by using these simple calculations are as follows:

(1) Not all prices follow the CPI trend. The method of calculation of the CPI itself has been modified at times. The base or reference year for the CPI has also been changed. We can show our students how to convert between different base-year values.
(2) Any "index number" can be misleading and may not apply to any particular situation. The relevance of an index number to you will depend on the components included in its calculation and whether you trade in those items.
(3) The "structure of prices" becomes distorted as inflation proceeds.

[Note: The exchange ratio between slide rule and calculator in 2011 was approximately 20, but in 1964 was "undefined"!]

6.3   The Language of Inflation

Remembering that "words have meanings," consider the following argument. Most, if not all, engineering economy books will say something along the lines of "inflation and deflation are terms that describe changes in price levels in an economy" (Thuesen and Fabrycky 2001). The authors of these books typically maintain that inflation means "increased prices." This statement equates the cause (inflation) with the effect (price increases) by changing the meaning of the word inflation.

The original meaning of the term "inflation" in the classical tradition was an increased supply of <u>money</u>. This increased supply of money which can be achieved by various methods that will be described later in this section might cause increasing prices, but other factors would also influence resulting prices.

The inversion in meaning of inflation has been explained by Greaves (1974):

> The currently popular fashion of defining inflation by one of its effects, higher prices, tends to conceal from the public the other effects of an increase in the quantity of money whenever the resulting rise in prices is offset by a corresponding drop in prices due to an increase in production. The use of this definition thus weakens the opposition to further increases in the quantity of money by political fiat or manipulation and permits a still greater distortion of the economic structure before the inevitable readjustment period, popularly known as a recession or depression.

Money prices are simply exchange ratios between money and goods; therefore, many factors might affect these prices. Some of these factors could be the supply of goods, changes in demand for particular goods, or deficit financing by governments, as well as changes in the amount of money or credit available. Detailed explanations of these issues related to the theory of money and credit are really beyond the scope of this narrative. However, students should realize that the supply of money can be intentionally increased by those controlling the monetary system. At this point in the narrative, we remind students about Mises' (1996) regression theorem and Menger's (1994) theory of money to reinforce understanding that money cannot be arbitrarily defined by some authority. Money is an evolved institution, as maintained also by Hayek (2011).

As the previous example and common experience show, prices do not all rise simultaneously or proportionally. A so-called price index can be valid only to the extent that the prices of its constituents reflect the general state of the economic system. An index number is an artificial aggregate price. The Austrian theory maintains that no index is valid. The concept of a price level or a standard of living is "confused" (Mises 1996). All prices are merely past data and not measures of value in any sense, especially for predictions of the future. Prospective calculations are always speculations about future conditions.

Entire books in the Austrian tradition have been written about money and credit (Mises 1981). It would be impossible and unnecessary to include a complete theory of money here. The main concern is to relate quantities of money and price increases so that engineering students appreciate the complexity of the issue of inflation and, consequently, price controls to be considered later.

We can distinguish between "money proper" and "money substitutes" to give a sense of the problem. According to Mises:

> Money proper... includes the following: commodity money (gold coin), credit money (claims to money not readily redeemable) and fiat money (money solely by reason of law) when commonly used as media of exchange. It does *not* include the following: token money (minor coins), money-certificates (redeemable claims to money) and such fiduciary media (q.v.) as bank notes and deposits against which the monetary reserves are less than one hundred percent.(Greaves 1974)
>   Money-substitutes include token money (minor coins), money-certificates (issuer maintains 100 % reserves in money proper) and fiduciary media (issuer maintains less than 100 % reserves in money proper). Fiduciary media in turn include both banknotes and bank deposits subject to check or immediate withdrawal. Money-substitutes serve all the purposes of money proper. They are part of money in the broader sense and a factor in the consideration of all catallactic problems as well as those affecting the money relation (q.v.). (Greaves 1974)

So, "money" (in the general sense) includes both "money proper" and "money substitutes." This is "the sense in which the term 'money' is used in discussions of the problems of catallactics...." (Greaves 1974) and, therefore, is the meaning of "money" when we deal with inflation and rising prices.

6.4 Methods and Episodes of Devaluation

Essentially, inflation devalues the money. How can the money supply be inflated? One possibility would be counterfeiting, which is illegal; counterfeiters are criminals. But, the "new" money goes directly to the counterfeiters who benefit immediately by having more purchasing power unless caught. So far as the general economic system is concerned, counterfeiting on this scale is of no consequence. Counterfeiting could not cause a general rise in prices.

There could not be a general rise in prices without a massive increase in the supply of money. And, further, there must be a continuing increase in the quantity of money to permit continuing price increases. Austrian theory says that to maintain this apparent stimulating effect would require an increasing rate of increase in the money supply. This would eventually lead to runaway inflation and a "crackup boom" (Mises 1996).

At this point, we should introduce Hayek's (2009) molasses analogy which will be considered in greater detail in Chap. 7. The new money progresses outward from the original recipients and spreads throughout the economy, and the effect (benefit) is reduced until the later recipients actually fall below the new price level and have actually lost purchasing power. Unless the rate of money creation increases, the inflation will lose the desired effect.

It would be instructive for students to see various ways that historic devaluations have been caused. To do this, we can begin with a statement of Gresham's Law; in short, "bad money drives out good money." For a more detailed description of Gresham's Law, see Greaves (1974).

Historically, devaluation of the currency called for debasing commodity money. Methods included clipping edges of gold and silver coins, and the response to protect the value of coins would have been milling and weighing the coins. Then, to devalue the coins, the commodity money would be diluted with base metals. The response to that would be assaying methods, such as testing gold coins with an acid. Merchants would be required to weigh and assay coins to assure their value in exchange. We could use this experience to introduce transaction costs.

Eventually, legal counterfeiting, that is, the printing of paper money, evolved from paper notes "payable on demand" (in gold or silver coins) to Federal Reserve "notes" based on the faith and credit of the federal government. I show the class facsimile copies of gold and silver demand deposit notes and a Federal Reserve note. The demand deposit notes clearly state "payable on demand" in gold or silver coins that are on deposit.

We can then examine the newest Federal Reserve paper money and consider the great effort taken to prevent illegal counterfeiting to make this money "safer." Contrast this with the potentially massive, legal counterfeiting of the Federal Reserve.

The example of hyperinflation that I use in class comes from Germany, 1923, to demonstrate what can and did happen in a highly civilized, cultured country—the land of Goethe and Beethoven. Sowell (2000) called that German hyperinflation the most famous inflation of the twentieth century and included enough data to allow students to make some estimates of rates of inflation caused by the printing of massive amounts of marks. For example, in October 1923, prices rose 41 % per day (Sowell 2000). Students should know that this inflation was intentional in response to reparation payments demanded by the Allies after World War I.

Other documented episodes of inflation that students would not know about include the American Revolution (1776–1780) when, over those four years, inflation averaged 150 % per year (Churchill 1999). The Confederate States of America (1861–1865) had an average inflation rate of more than 208 % per year (McPherson 1988).

Hazlitt (1996) includes inflation rates for numerous countries around the world and asks the reader to ponder the chaos and suffering caused by these rates of depreciation of the value of currencies. His conclusion is that the "one lesson" of economics has never been learned by governments. Having assigned this book as summer reading, I ask students on an examination to restate the lesson: "…the whole of economics can be reduced to a single lesson, and that lesson can be reduced to a single sentence. *The art of economics consists in looking not merely at the immediate but at the longer effects of any act or policy; it consists in tracing the consequences of that policy not merely for one group but for all groups.*" (Hazlitt 1996)

Argument in favor of a metallic (gold) standard money is basically to restrict government expansion of the money supply. Since the supply of gold is limited and expensive to increase, having paper currency redeemable in gold would limit the amount of paper money that could be issued (Sowell 2000). Of course, a major discovery of gold could cause an inflation, as with the Spanish removal of gold from the Americas to Europe. However, such an event would not be of the magnitude of inflations caused by printing paper.

6.5 Estimating Devaluation Effects

At this point, we can proceed to define a price index as a ratio of current price to a base price. The price index should match the factors involved in a particular study as well as possible. This index will be used to adjust for monetary depreciation.

We can use the CPI because it is usually tabulated in textbooks. We can also find more current CPI data from government Web sites which will allow us to recalculate CPI values because of a change in the CPI base year. Then, we can use the CPI ratio to calculate an average inflation rate over some time period, basically using Eq. 3.1, again.

The introduction of inflation calculations causes us to identify three types of interest rates that we will need:

(1) "market interest rate" (i): assumed to include inflation, deals with "actual" dollars (this is the number of dollars you would have to count out at any future time)
(2) "inflation-free rate" (i'): an abstraction, the effects of inflation are removed, works with "constant" or "base-year" money units (e.g., dollars)
(3) "inflation rate" (f): the escalation rate or the percent increase in prices.

Since these are interest rates, they can be specified for any unit of time, usually we think in terms of yearly rates. Whenever any of the rates are used, we are simply guessing about the future.

The introduction of inflation also causes us to work with two types of money definitions. We define "actual dollars" as "out of pocket", current, escalated, or inflated dollars. These can be imagined as physically real paper dollars at any point on a cash flow time line. And, we define "constant dollars" as real, deflated, zero-date, or base-year dollars having the effects of inflation on purchasing power removed. These real dollars are abstractions.

When we want to convert between real and actual dollar amounts, we can use the following formula, noting that typically a greater number of actual dollars will be needed to maintain a given constant dollar value.

$$(\text{actual dollar amount}) = (\text{constant dollar amount}) * (1 + f)^n$$

Emphasize that this is merely another application of Eq. 3.1. It can be rewritten in terms of some price index, such as the CPI, as follows:

$$(1 + f)^n = \text{CPI}@n/\text{CPI}@0$$

What this means is that we can calculate in hindsight a so-called price inflation rate, based on any index number. However, our economy studies are for the future. The index ratio yields a geometric average inflation rate, f, because of the compounding effect of inflation.

For example, using the 25-year period 1967–1992, we can calculate, in retrospect, the apparent average annual inflation rate as follows:

$$(1 + f)^{25} = 420.3/100 = 4.203$$

The average inflation rate, f, seems to be 5.91 % annually. For example, if someone retired in 1967 with an annual income of $10,000 which was considered to be a good salary at that time, the "constant dollar" income in 1992 would have been only $10,000/4.203 = $2380 (base-year dollars). This simple example demonstrates the destructive power of even a "moderate" inflationary effect. Other reference dates could be used to show the same effect. Consequently, one can easily imagine the total devaluation caused by a 150 % annual or 41 % daily inflationary episode.

Example 6.2 Suppose that we have $100,000 (in 1967, today), so we encumber a purchase that will require a "balloon" payment of $500,000 "constant," base-year dollars at the end of 1992. The contract was written in terms of gold dollars! The question for students to answer is as follows: "What market rate of interest must we realize to be able to meet this obligation?"

This question causes us to apply two "domains" for a solution. In the "constant dollar domain," we can find that because

$$\$500,000 = \$100,000(1 + i')^{25}$$

the inflation free interest rate, i', is 6.65 % annually.

Then, converting (inflating) the constant dollar payment (in 1992) to the "actual dollar domain" using a version of Eq. 3.1 gives:

$$F = \$500,000(1 + 0.0591)^{25} = \$2,101,500, \text{ or using the CPI ratio of 4.203,}$$
$$F = \$500,000(4.203) = \$2,101,500.$$

Then, in the actual dollar domain, we can find the required market interest rate from:

$$F = P(1 + i)^{25}$$
$$(1 + i)^{25} = \$2,101,500/\$100,000$$

so that the market rate, i, must average 12.954 % annually.

From this calculation, we clearly see that the market rate does not equal the inflation-free rate plus the inflation rate. This is not the result that students expect. We can ask that students follow the mathematical proof based on the interest rate factor formulas that is found in most textbooks to learn that

$$i = i' + f + (i)' * (f)$$

We should also emphasize that the conclusion of this example has still not accounted for taxation. We could ask students what they expect that effect to be.

To get them to think, I often tell students that it would be nice if taxes were levied on a constant dollar basis, but the taxing authorities would neither understand nor agree!

6.6 Accounting for Inflation in Two Domains

Textbooks typically demonstrate that a cash flow problem can be worked numerically in either the "constant dollar" or the "actual dollar" domain, as long as proper accounting of i, i', and f is made. This process is included in Chap. 5 of Thuesen and Fabrycky (2001), my usual course textbook. Example problems from Chap. 6 of Au and Au (1992) have served as good starting points for in-class demonstrations. I require students to solve some problems by both analysis methods. They are reminded to keep these problems in mind because they will be the basis of a more complex analysis after the effect of taxation is added.

After students have seen the effect of inflation, they should proceed to Chap. 7 on information and price signals for a more complete understanding of the importance of the interaction of inflation and price signals, i.e., the distortion of the price structure, a major Austrian insight.

References

Au T, Au T (1992) Engineering economics for capital investment analysis, 2nd edn. Prentice Hall, Englewood Cliffs
Churchill W (1999) The great republic. Random House, New York
Friedman M (1994) Money mischief. Harcourt Brace, New York
Greaves P (1974) Mises made easier. Free Market Books, Dobbs Ferry
Harford T (2011) Adapt. Farrar, Straus, and Giroux, New York

Hayek F (2009) A tiger by the tail. Ludwig von Mises Institute, Auburn

Hayek F (2011) The constitution of liberty. Chicago

Hazlitt H (1996) Economics in one lesson. Laissez Faire Books, San Francisco

McPherson J (1988) Battle cry of freedom. Oxford

Menger C (1994) Principles of economics. Libertarian Press, Grove City

Mises L (1981) The theory of money and credit. Liberty Fund, Indianapolis

Mises L (1996) Human action, 4th rev edn. The Foundation for Economic Education, Irvington-on-Hudson

Sowell T (2000) Basic economics. Basic Books, New York

Thuesen G, Fabrycky W (2001) Engineering economy, 9th edn. Prentice Hall, Upper Saddle River

Chapter 7
Information: Interfering with Price Signals

The subject of this section is primarily the transfer of information among economic actors. A closely related concern is distortion of the information caused by various interventions. We note that we are always trying to plan action into the future, while the future is always uncertain, or unknowable. Even prospective calculations are based on data from the immediate past. The circumstances surrounding any proposed action are affected by the passage of "real time" and the conditions of "ignorance." The themes of *time* and *ignorance* are borrowed freely from the book by O'Driscoll and Rizzo (1996).

7.1 Information Theory and the Market

In a recent book, Gilder (2013) presents an analogy between information theory and the market economy. He maintains that a "low-noise" carrier is necessary so that the information that he calls "surprises" can be identified. Government should create stable, low-noise systems in the economy so that entrepreneurs can safely innovate. This requirement is closely related to Hayek's description of the use of dispersed knowledge (1977). This means that regulations, changing rules, monetary manipulations, and taxes are excessive noise in the "carrier." These ideas are compatible with Austrian theory as described by O'Driscoll and Rizzo (1996).

Gilder's "information" theory basically concerns the entrepreneur, or innovator, looking ahead, predicting future conditions based on specific, individual knowledge. As he says, "knowledge is about the past. Entrepreneurship is about the future" (p. 108). This statement agrees with Austrian theory that price data are from the past, it is about history, not the future (Mises 1996).

Gilder's description of entrepreneurial activity is also compatible with Hayek's description of the "use of knowledge in society." That is, knowledge is diffused among many individuals so that there must be some mechanism available to coordinate activities. In his case, this is the price system.

© Springer International Publishing Switzerland 2015
R.A. Chadderton, *Purposeful Engineering Economics*,
DOI 10.1007/978-3-319-18848-5_7

O'Driscoll and Rizzo (1996) note "...there is a form of uncertainty that cannot be eradicated by further knowledge. This is the time-dependent aspect of genuine uncertainty." As "real time" passes data change and "insight" needed for successful prediction of future action must adjust.

Thus, we can say that a producer (entrepreneur) makes prospective calculations based on personal knowledge, or "insight", and then retrospective calculations based on information (prices) passed to him from consumers. In retrospect, the price data passing from consumers to producers (or, to entrepreneurs) can either confirm or contradict that foresight (anticipation). Profit or loss, if calculated properly based on reliable information, can guide future activity to better outcomes.

Gilder and Hayek could agree that the "noise" or distortions caused by government interventions in terms of regulations, changing rules, monetary manipulations, or taxes, which Gilder calls excessive noise in the carrier and Hayek calls price distortion, are causes of confusion to the market economy. This confusion leads to mal-investment of scarce resources.

7.2 The Evolution of Social Institutions

Another theme that we should keep before the students is the evolved nature of social institutions, including money, by which we mean that they were not planned by anyone (Hayek 2011). Menger's theory of money (1994) describes the evolution of money. Mises' regression theorem (1996) traces the use-value of money back to a time when the monetary good (for example, gold) was only used for non-monetary purposes. He considered the evolution from commodity use to monetary use as critical and concluded that a government could not simply define a money of its choice that would be accepted for general use in exchanges (i.e., a legal tender law).

In a classic article that students should read, Read (1992) explained why "no one knows how to make a pencil"—and he meant a wood pencil! The basic reason that pencils can be made, distributed, and sold is that prices coordinate the actions of innumerable unknown participants who do not know others involved in the process. The "price structure" in terms of the evolved institution of money can be seen as the guide to coordinate all the activities needed to make pencils. Rational individuals need only to follow reliable price signals. Students need to understand the extent and complexity of this coordination as best described by Hayek (1977).

7.3 Interference by Inflation

Since prices coordinate the multitude of individual actions, interferences with the price structure can be seen to cause dis-coordination. I want students to realize the significance of this distortion of the structure of prices caused by four activities, namely inflation, taxation, price controls, and "public" projects.

The major interference that I want the students to encounter is the manipulation of the monetary unit by the monetary authorities. In Chap. 6, we saw that many eminent economists maintain that inflation is a deliberate policy of the monetary authorities. The tendency is to devalue the monetary unit, a process that the classical economists termed inflation. Currently, the word "inflation" is applied incorrectly to the phenomenon of rising prices. Our objective is now to consider the effect of inflation on information or price signals.

Inflation affects different components of a cash flow to different degrees; it affects different people differently, as well. Inflation in the classical sense is the injection of new money (or credit) into the economy by the monetary authority. This new money is not spread uniformly throughout the economy, but rather goes immediately to select beneficiaries who provide goods or services to governmental authorities.

Hayek (2009) has described this injection as a pouring of molasses into a container. I see the container as a graduated cylinder—the molasses is poured vertically at the center of the cylinder's open end. The rate of input is held constant so that the average surface height in the cylinder rises at a constant rate as well. The rate of rise can be calculated from a basic fluid conservation principle. However, the molasses will mound in the center and spread outward to the wall of the cylinder.

Those who receive the new money first "win" because they get the extra new money and its purchasing power before the average height (or average "price level") rises. Those who get the new money last have "lost" because the average price level has risen before they get a share of the new money which has already lost purchasing power as indicated by the higher, and rising, average price level.

Hayek's observation falsifies simplistic solutions to inflation, such as dividing all prices by some factor to reduce the price level. Defining a "new dollar" as being worth ten "old dollars" would not eliminate the distortion of the price structure as proponents of the scheme might argue.

If the rate of injection of new money remains constant (like a constant rate of fluid flow), then the rate of rise in the cylinder (the "price level") can be calculated and anticipated. This anticipated rate of inflation would have limited effect on individual actors because their plans could be adjusted to account for the anticipated rate of price increase. Therefore, the rate of inflation must be continually increased to create a constant state of "surprise" that might keep the "boom" economy going, or in other words to "prop up" the mal-investments caused by inflationary expectations. We can point to Mises (1996) for a more complete explanation of this effect.

For an individual, the result of the inflation policy is to create two types of prices that we know as "nominal" (actual or today's prices) and "real" (base year or constant dollar prices). And, because inflation affects different items to different degrees, the price structure in real (base year) terms would have been distorted into actual (nominal) terms. Decisions based on actual dollar amounts might not have been appropriate if prices had been adjusted to constant dollar amounts. Price signals had been destroyed. Essentially, such a misdirection of resources can be called "mal-investment" which would have to be liquidated when the inflation slows. The correction process is commonly called a recession. Note the irony that the monetary authority can take credit for "fighting" its own inflationary policy and for "fighting" the resulting recession!

7.4 Interference by Taxation

We must also consider the effect superimposed on this inflationary environment by taxation. Taxes, of course, are based on actual (current, nominal) amounts. This introduces the concepts of before-tax cash flow (BTCF) versus after-tax cash flow (ATCF). Since our available purchasing power is the after-tax residual, we need to be prepared to convert from BTCF to ATCF. The process will be to assume that our initial cash flow is stated in actual dollar amounts. An applicable tax rate (which might be the anticipated rate) can be applied to the actual dollar BTCF, taxes subtracted and the actual dollar ATCF calculated. If we want, we could then convert back to a real or constant dollar ATCF. Using an assumed, appropriate discount rate for either the actual or constant domain, a decision variable such as net present value (NPV) of the ATCF can be calculated for decision purposes.

Students must recognize the significance for decision making of the combined effects of inflation and taxation. Examples to follow (Chap. 9) will show that including some, all, or none of these effects will alter the ultimate decision, possibly reversing an investment decision.

7.5 Interference by Price Control

Two additional causes of price distortions that I want students to understand are the effects of price controls, such as rent control or minimum wage laws, and the redistribution of purchasing power due to so-called public investments. Price controls will be considered in this section. Chapter 8 will be dedicated to the subject of so-called public investments.

We can maintain that the supply of essentially anything tends to increase with rising price, while the demand for it tends to decrease with rising price. This tendency applies to labor or leisure as well as physical products. We can visualize a supply and demand graph as two intersecting straight lines that form an X and label the price at the point of intersection as P. The amount of product both supplied and demanded at price P can be called N. For our purposes, the point P specifies the price that brings supply and demand into agreement; the market clearing price is P, and the transaction amount is N. I would omit numerical values to emphasize the uncertainty of such a graph.

7.5.1 Minimum Price Control

We now want to consider the effect of a controlled price C, representing a minimum wage law. First note that if C is less than P, it is ineffective because buyers are already willing to pay suppliers of such work more than the permissible minimum wage. This is a perfect situation for politicians; it gives the appearance of action without affecting this particular labor market!

But, consider the case that receives so much publicity in the media, when C is more than P. Such a case (an arbitrarily raised allowable minimum wage rate) has clearly distorted market price signals and sent false information. It has driven marginal buyers (employers) out of the market because they can no longer afford to pay so much for the level of productivity of the workers while drawing more suppliers (workers willing to work at the higher wage) into the market. The reduced number of transactions at C represents unemployment caused by the new requirement.

The law that raised the minimum allowable wage rate has surely benefitted those workers who can keep their jobs! That much is seen and publicized by politicians and journalists alike. What is not seen or ever mentioned is the forced unemployment (excess demand for these jobs) created by the attraction of additional workers into the workforce at the higher wage rate. This result can be called institutional unemployment (Greaves 1974). Such a policy clearly penalized the less skilled workers, particularly the young.

Williams (2008), Sowell (2000, 2009), and others have labeled minimum wage laws racist because they most affect young black men who have limited skills. I ask my students to learn about Walter Williams and Thomas Sowell to see why their perspectives are particularly appropriate in today's political climate. Any politician who promises to "raise the minimum wage to benefit thousands" should never be elected by the thousands who are hurt by that promise if they understand the result of the price control.

7.5.2 Maximum Price Control

Next, consider the result of a ceiling, or maximum, price control. We all know the stories of urban decay and the accompanying issues of crime and slumlords. These are the visible manifestations of a problem. I want students to have some level of appreciation and understanding of the unseen causes of these public problems, the causes that politicians and journalists do not acknowledge. The problem of "rent control" is discussed in many places; I choose to cite Thomas Sowell (2000, 2009).

Specifically, what does "rent control" actually do? "In short, with housing as with other things, less is supplied at a lower price than at a higher price—less both quantitatively and qualitatively. Polls of economists have found virtually unanimous agreement that declines in product quantity and quality are the usual effects of price controls in general. Of course, there are not enough economists in the entire country for their votes to matter very much to politicians, who know that there are always more tenants than landlords and more people who do not understand economics than there are who do. Politically, rent control is often a big success; however, many serious economic and social problems it creates." (Sowell 2000).

The result of an <u>effective</u> controlled rent level when C is less than P is an increased demand for a reduced supply of rental space relative to the market clearing amount and price, P. An <u>ineffective</u> rent control price would be indicated by C higher than P. The usual governmental response to such shortage caused by an effective price control is some form of rationing, such as was seen during the "gas crisis" of the 1970s. If followed to its logical conclusion, this scheme of interference would lead to a situation where all prices would have to be controlled, a condition known as German socialism (Greaves 1974). I want students to be aware of such a limit to bureaucratic intervention in the economy.

7.6 Summary

Additional market interventions that have distorted the information transmitted by prices include minimums for agricultural products and maximums for gasoline. Many economists, including Rothbard (2004) as well as Sowell (2000, 2009) and Williams (2008), have elaborated on the effects of various price controls.

We can conclude that any effective price control distorts information that both suppliers and demanders need and creates perverse, counterproductive incentives. To appreciate fully these incentive distortions requires an understanding of the unseen, long-term consequences for the general public, not just the short-term benefits to a particular segment of the population. The classic description of this process which has occurred over many decades is Hazlitt's book *Economics in One Lesson* (1996) that I have assigned as outside reading for my classes.

The cumulative effect of inflation, price control, public spending, taxation, and regulation is the redistribution of wealth and creation of debt. An appreciation of the individualist, Austrian analysis of these effects would help students contradict the Keynesian fallacy that the national debt does not matter "because we owe it to ourselves." They would understand that that debt is owed by some individuals to other individuals. It is not a zero-sum situation. For an exhaustive explanation of the entire process of price control, inflation, taxation, mal-investment, and other secondary effects, we can refer students to Hayek (2011).

References

Gilder G (2013) Knowledge and power. Regnery, Washington, DC
Greaves P (1974) Mises made easier. Free Market Books, Dobbs Ferry
Hayek F (1977) The use of knowledge in society. Institute for Humane Studies, Menlo Park
Hayek F (2009) A tiger by the tail. Ludwig von Mises Institute, Auburn

Hayek F (2011) The constitution of liberty. Chicago

Hazlitt H (1996) Economics in one lesson. Laissez Faire Books, San Francisco

Menger C (1994) Principles of economics. Libertarian Press, Grove City

Mises L (1996) Human action, 4th edn. The Foundation for Economic Education, Irvington-on-Hudson (revised edn)

O'Driscoll G, Rizzo M (1996) The economics of time and ignorance. Routledge, New York

Read L (1992) I, Pencil. Imprimis 21(6):1–3

Rothbard M (2004) Man, economy, state. Ludwig von Mises Institute, Auburn

Sowell T (2000) Basic economics. Basic Books, New York

Sowell T (2009) Applied economics. Basic Books, New York

Williams W (2008) Liberty versus the tyranny of socialism. Hoover, Stanford

Chapter 8
Public Projects (Benefit–Cost Analysis)

The material in this section is by its nature political. The emphasis will be on benefit–cost (B/C) analysis of government activities, or more specifically of projects. Arnn (2012) is particularly relevant to this material. This section begins with a philosophical discussion and ends with a mathematical example.

8.1 "Philosophical" Introduction

The country in which we live is much different from the one envisioned in our founding documents. This is evidenced by several recent books (Arnn 2012; McClanahan 2012; Sandefur 2014; Epstein 2014) and many references to events of history.

As Arnn (2012) describes the situation, *"The government now approaches half the size of the economy as a whole....Our retirements, our health, and the relations inside our families are now the business of the federal government. Each business, large and small, is also under its purview. It is so pervasive that it seems to be the only way for the society to work."* And yet, a recent poll showed that only 17 percent of the people think that this government works with the "consent of the governed" (Arnn 2012). There is a clear disconnect between the government and the citizens of our country. This is a main reason that I remind students that the government is not the country.

The costs of government as described by Gwartney et al. (2010) include the loss of (unknowable) private sector output, the costs of enforcement and compliance (regulations), and the costs of price distortions (exchanges that do not happen) in addition to the burdens of taxation. In this section, the validity of many of the activities of governments is questioned in part because of the extensive price distortions and dis-coordination that result. Ultimately, engineers are required to use benefit–cost (B/C) analysis as the appropriate method to evaluate individual public projects.

© Springer International Publishing Switzerland 2015
R.A. Chadderton, *Purposeful Engineering Economics*,
DOI 10.1007/978-3-319-18848-5_8

Ironically, the price data upon which the planners must rely for *B/C* analyses are not only inaccessible (the idea of dispersed knowledge) but also incorrect (distorted) because of massive effects of previous government actions and the resulting costs. These issues were considered in Chap. 7. Nevertheless, *B/C* analysis is commonly used to justify public works. One objective of our discussion should be to argue for a minimization of intervention to preserve the relevant price data. Causes of distortion previously introduced (Chap. 7) were as follows: inflation, price controls, regulations, and taxation.

The Founders favored local to centralized government; constitutional guarantees did not include any mention of "education, health, retirement, welfare, or any of the hundreds or perhaps thousands of areas of policy in which the federal government now operates" (Arnn 2012).

8.2 Government Incentives and Functions

An evaluation of public activities is truthfully an attempt to evaluate government activities. But, government is an imaginary aggregate, not an acting individual or unit considered by the Austrian economists. Government-sponsored projects are promoted by specific compartments (units) of a more general government; the unit would have its own agenda and priorities, essentially ranked preferences.

Students should be aware that politicians have much different incentives compared to private individuals. Politicians benefit from visible, seen activities that have immediate short-term concentrated effects. Actions that have only long-term, unseen benefits for the general public can be postponed to be dealt with by a future politician. Sowell (2009) compared infrastructure maintenance that can be ignored by current politicians against facilities offering ribbon-cutting ceremonies that have immediate benefit to elected officials to contrast these different incentives. As stated by Sowell (2009), the "real question is not which policy or system [government or market] would work best ideally, but which has in fact produced better results with far from ideal human beings."

The discussion of "what is seen," the visible results of government's activities, and "what is unseen," the unknown private investments that were not made because of the government's appropriations, can be supported by references to Bastiat (2011), Hazlitt (1996), or Mises (1996).

When I reach this point in my class, I try to relate the topic of *B/C* analysis to current constitutional issues. The students will have all been previously given pocket-size copies of the Declaration of Independence and US Constitution (2002). They are urged to search these booklets for support of economic calculations to follow.

When I revised this material on *B/C* in November 2012, more than 1000 federal programs were facing budget cuts. How many of these programs would be considered appropriate governmental undertakings by the Founders, or the Austrian economists, is a question that should be addressed. The answer would be "very few."

Legitimate, essential functions of government are national defense, police, and courts. Thus, the proper role of government in a free society is very limited, to protect against "criminals foreign and domestic." One good example of this argument is an essay by Williams (2000).

Yet, there is no doubt that the federal government undertakes innumerable activities not mentioned in the founding documents, including minimum wages; price supports; education; and now health care. We should also note the difficulty caused by too many Americans who now look to the government to solve their every problem. Most citizens have become trapped in the system of redistribution and benefits; they cannot afford to opt out of federal social security and healthcare programs.

The welfare state concept is not new; it is the offspring of Bismarck's socialist-autocratic system that has been both admired and emulated during the past century (Lane 1984). The drift of the US mixed economy toward the fascist version of socialism, that is, an economy comprised of seemingly private institutions being told what to do by administrative agencies, can be documented by reference to descriptions in Greaves (1974). This is a fundamentally flawed progression; engineers may not be able to stop it, but we can understand why events frequently turn out badly. A good engineering example is water resource projects that never repay investment (Hirschleifer et al. 1970).

Misinterpretation of the so-called "general welfare" clause of the US Constitution has opened the door to numerous special interest promoters and manipulators. The "general welfare" originally referred to the nation as a whole, not to any individual's personal welfare. That is, "…the "general welfare" meant legislation that benefited each State and defended the liberties, "religion, sovereignty, [and] trade" of the several States" (McClanahan 2012). The Founders did not intend for the general government to act directly on the citizens.

For example, Martin Van Buren (1837) maintained that "All communities are apt to look to government for too much….But this ought not to be. The framers of our excellent Constitution…acted at the time on a sounder principle. They wisely judged that the less government interferes with private pursuits the better for the general prosperity" (quoted by Carson 1985). We note that this could be a summary of the Williams' article (2000) written for today.

8.3 Limitations to Government Activities

At this point in my class, I ask the students to find reference to health care in their copies of the US Constitution that were given out in class. A survey of those CEE students in 2014 showed that most of those who responded cited the general welfare clause as the justification for the recent healthcare law. This can lead to a consideration of problems created by government controls, because I want students to be aware of the consequences of reliance on bureaucratic management, as elaborated by the Austrian school economists.

Among the severe limitations that are coincident with bureaucratic management of so many public functions are:

– the one who pays is different from the one who receives, the issue is incentives;
– "perverse incentives" created by a system of diffused costs versus concentrated benefits (the beneficiaries have an incentive to organize and promote their program while the multitude of taxpayers do not);
– This creates another kind of B/C problem if one appreciates irony: the beneficiary $B/C \gg 1.0$, while the taxpayer $B/C \ll 1.0$ leading to "rational ignorance" on the part of the general taxpaying public!
– governmental bureaucracies are, by necessity, inefficient; daily reports of waste are evidence (Mises 2007); and
– there will always be unanticipated consequences.

Unanticipated consequences should, but under centralized management will not, fall on the "managers" as would happen in the private sphere. Therefore, corrective incentives do not develop; "business" continues as usual. [Could one not cite the recent scandal at the VA?]

The impossibility to foresee the future; all prospective planning is speculation, best done by those who have specific knowledge.

– Problems of "use of knowledge" by centralized planners versus individuals; the role of prices: "*Fundamentally, in a system in which the knowledge of the relevant facts is dispersed among many people, prices can act to co-ordinate the separate actions of different people in the same way as subjective values help the individual to co-ordinate the parts of his plan.*" (Hayek 1977).
– Thuesen and Fabrycky (2001) note that even the one who receives benefits does not know the value or the cost.
– As Mises (2005) noted in reference to the contradictory promises of political parties; "Of course, one can simultaneously promise city-dwellers cheaper bread and farmers higher prices for grain, but one cannot keep both promises at the same time." [One might add, "without massive transfers of wealth from third parties"!]

In short, the more distorted prices become because of the numerous previous interventions of government, the more out of balance economic activities become. A result is that the general level of well-being is forced downward.

We see in the never-ending promises and ever-expanding realm of governmental activities the need of every branch of government for more resources. Therefore, we should consider the ways that government can finance these so-called public activities. The government cannot give something to one person before it takes that something from some other person. The financing methods available are limited to three, namely taxation, accumulation of debt, or devaluation of money.

8.4 The Method of *B/C* Analysis

The general objective of *B/C* analysis is to compare an independent project (the least "costly" choice) against the null (do nothing) alternative. Then, an incremental analysis (d*B*/d*C*) is used to make successive pairwise comparisons among a set of mutually exclusive alternatives (Au and Au 1992). Our assumption will be that the projects to be evaluated by *B/C* are only those that are commonly assumed to be incapable of being privately provided, such as flood control or transportation systems. However, we should cite counter claims. For example, railroads and highways could have been financed privately (Lane 1984). We can also refer to Kazmann (1972) on the ineffectiveness of flood control at great and increasing cost.

Students must know to avoid double-counting of benefits for multi-purpose alternatives. Our examples will assume honest estimates of costs and benefits. Then, one could select the "best" alternative for a particular project based on incremental *B/C* analysis.

However, Meiners (2014) noted that "government *C-B* practitioners now make up prices" for non-market items. One example was the "price" of carbon dioxide emissions used to justify more stringent regulations on coal and energy use. Meiners maintained that this is not good economics; that *C-B* analysis using made-up prices is central planning, not science. His conclusion is that economists should be more modest (Meiners 2014).

We must also emphasize that *B/C* analysis cannot determine whether the selected alternative was the most crucial of all projects that could be undertaken. For example, to choose among an airport, a highway, or a water supply project, which are not mutually exclusive, political decisions would be required. Competing "units" and "information" problems would be involved. Further, an overall budget limitation might eliminate an optimal alternative from the incremental analysis.

8.5 A Typical Five-Step Procedure for Incremental *B/C* Analysis

1. Calculate present values of costs and benefits separately using an appropriate interest rate (MARR or *i*).
2. Arrange alternatives in the order of increasing PV of "costs"; [PVC(*i*)].
3. Compare the lowest cost alternative against "do nothing" alternative.
4. Progress through the alternatives pairwise, increasing [PVC(*i*)] until all but one alternative have been eliminated.
5. As a check, the last remaining alternative, the selected project, should also have the maximum net present value or NPV(*i*). [This conclusion comes from Dr. Au who maintained that it would be unnecessary to use *B-C* analysis instead of NPV methods "if you can influence the rest of the world." (Au and Au 1992).]

Students should have learned the conditions that require an annual equivalent (AE) decision method, but also know that AE's for different projects of different durations can be reduced to the appropriate PV's. I emphasize PV calculations to be readily available for Dr. Au's suggestion that maximized NPV would select the same alternative as correctly done incremental *B/C* analysis.

Whether one represents costs and benefits separately as PV or AE, numerical evaluations require the use of an appropriate discount rate. The question becomes how to select an interest or discount rate for governmental activities.

8.6 Choice of Interest Rate for *B/C* Analysis

We must accept the fact that government units undertake many activities. Funding those activities requires financial resources that have time value. The question then is how we should account for the time value of proposed project cash flows appropriately. Discussion of the so-called social discount rate can be found in various engineering economy textbooks, e.g., Au and Au (1992) or Park (2013).

One possible argument is that, because government can borrow at a lower rate of interest than can private businesses, a lower discount rate compared to market rates should be used. As our basic formula (Eq. 3.1) easily shows, such a case discounts future values at a lower rate, giving distant future values a higher present value than an investor would calculate for a private transaction.

An alternative argument is that the rate of discount for *B/C* studies of government projects should reflect true time preference by using a market rate. The discount rate for government projects should reflect the foregone opportunity of unknown private investments due to the incidence of taxes. One good reference here is Hirshleifer et al. (1970).

The use of a discount rate that is too low will overstate value of a project relative to the private sector and increase the optimum size of a project. Thus, "Use of a low discount rate is favorable to the construction of public works projects and the interests which profit by project construction, but excessive diversion of resources to the public sector is detrimental to economic and even social efficiency and thus the long-run welfare of the nation. Solutions to pressing current needs may have to be sacrificed for the benefit of those living in the distant future. It is not possible to defend any exact discount rate for use in government planning dogmatically, but too low a rate definitely has serious adverse consequences to national economic growth." (James and Lee 1971). This statement implies sustainability issues and will be reconsidered later (Chap. 12).

Arrow (1972) also concluded that "argument for a social rate of time preference distinct from individual rates is basically a matter of value judgment. Its validity and its importance, if valid, are both subject to considerable dispute."

Hirshleifer et al. (1970) probably has as strong an Austrian flavor as any engineering book that I know. They prefer private to government activities in natural resources; they prefer local to federal intervention; they note the problem of information for centralization; they note government (water supply) projects financed at 2 % annual interest while suggesting that 10 % discounting would be appropriate for the same reasons as noted by James and Lee (1971).

Hirshleifer et al. (1970) also note the consequences of over-investment in projects that are too large and are made prematurely. These mistakes are described with respect to railroad construction, both in the USA and internationally (Chap. 11). In general, railroads were built too large and too soon (Rothbard 1978; Wolmar 2010).

8.7 A Numerical *B/C* Example

These effects (overbuilding especially) are easily demonstrated in the following example *B/C* problem suggested by portions of an example by Au and Au (1992).

The required calculation formula is simple, namely:

$$BC(i) = \text{Equivalent Benefits}(i) / \text{Equivalent Costs}(i)$$

Example 8.1 Assume four, mutually exclusive projects each having an expected useful life of 40 years with no salvage value at year 40 and a discount rate of 7 % annual.

Alternative (j)	Co(initial)	C(O&M) annual	B(annual benefits)
1	2.000	0.15	0.40
2	3.73	0.32	0.71
3	11.2	0.21	1.20
4	11.2	0.36	1.33

Note Tabulated values can be specified as millions of dollars or some other convenient unit; I sometimes call these "Villanova Units"

Step 1   Compute present values of costs and benefits for each project from the following equations:

$$CPV(j) = Co + C(O\&M)(P/A, 7\%, 40)$$

$$BPV(j) = B(P/A, 7\%, 40)$$

$$(P/A, 7\%, 40) = 13.3317$$

Step 2 Order the mutually exclusive projects in the order of increasing CPV(i):

Alternative (j)	CPV(j)	BPV(j)	dC	dB	dB/dC	Chose
0	0	0				
1	4.0	5.33	4.0	5.33	1.33	1
2	8.0	9.47	4.0	4.14	1.04	2
3	14.0	16.0	6.0	6.53	1.09	3
4	16.0	17.73	2.0	1.73	0.86	3

Step 3 has been followed, line by line, down the table making stepwise comparisons until all except Alternative 3 has been eliminated for $i = 7\%$.
Step 4 Calculate NPV for the four alternatives at 7 %.

Alternative (j)	BPV(j) − CPV(j) = NPV(j)	Total BPV(j)/total CPV(j)
1	5.33 − 4.00 = 1.33	1.33
2	9.47 − 8.0 = 1.47	1.18
3	16.0 − 14.0 = 2.0	1.14
4	17.73 − 16.0 = 1.73	1.11

Thus, we should choose Alternative 3 based on maximum NPV at 7 %. This is the
same result as given by the incremental benefit–cost ratio. However, it does not
maximize the total benefit/cost ratio. That was achieved for Alternative 1.

8.8 Gaming the Basic Example

Some further amplifications of this problem can be made by asking students for
the internal rate of return for Alternative 3 (answer = ~8.5 %). Another question to
address to students would be to ask what would happen to the decision if the true
"opportunity cost" of capital is more than 7 %? For example, I have them work out
this problem for the next class meeting at 9 %. (answer = Alternative 1).

Yet another question to ask is what interest rate would cause them to select the
"do nothing" alternative (that is, none of the projects would be justifiable). The
answer to this question is approximately 13 %.

This exercise demonstrates the effect of using an interest rate that is too low,
leading to a project that is too large (expensive). It shows why government agencies insist on using rates less than market rates ("Government gets preferred rate.")

To summarize these results, for a discount rate of 7 %, Alternative 3 (larger)
would be selected; for 9 %, Alternative 1 (smallest) would be selected; and, at
13 %, only the "do nothing" alternative could be justified.

A final interesting question would be: What is the rate of return for the "do
nothing" proposal when the MARR is 13 %? (Answer = 13 % which is shown by
an internal rate of return calculation).

References

Arrow K (1972) Criteria for social investment. In: Dorfman R, Dorfman N (eds) Economics of the environment, selected readings, 3rd edn. Norton, New York

Arnn L (2012) The founders' key. Thomas Nelson, Nashville

Au T, Au T (1992) Engineering economics for capital investment analysis, 2nd edn. Prentice Hall, Englewood Cliffs

Bastiat F (2011) That which is seen, and that which is not seen. Dodo Press, UK

Carson C (1985) A basic history of the United States, vol 3. American Textbook Comm, Wadley, Ala

Epstein R (2014) The classical liberal constitution. Harvard University Press, Cambridge

Greaves P (1974) Mises made easier. Free Market Books, Dobbs Ferry

Gwartney J, Stroup R, Lee D, Ferrarini T (2010) Common sense economics. St. Martin's Press, New York (revised edn)

Hayek F (1977) The use of knowledge in society. Institute for Humane Studies, Menlo Park

Hazlitt H (1996) Economics in one lesson. Laissez Faire Books, San Francisco

Hirshleifer J, DeHaven J, Milliman J (1970) Water supply economics, technology, and policy. University of Chicago Press, Chicago

James L, Lee R (1971) Economics of water resources planning. McGraw-Hill, New York

Kazmann R (1972) Modern hydrology, 2nd edn. Harper & Row, New York

Lane R (1984) The discovery of freedom. Laissez Faire, New York

McClanahan B (2012) The founding fathers' guide to the constitution. Regnery, Washington, D.C

Meiners R (2014) Glorification of cost–benefit analysis. PERCReports 33(1):12–13

Mises L (1996) Human action, 4th edn. The Foundation for Economic Education, Irvington-on-Hudson (revised edn)

Mises L (2005) Liberalism. Liberty Fund, Indianapolis

Mises L (2007) Bureaucracy. Liberty Fund, Indianapolis

Park C (2013) Fundamentals of engineering economics, 3rd edn. Pearson, Upper Saddle River

Rothbard M (1978) For a new liberty. Collier Macmillan, New York

Sandefur T (2014) The conscience of the constitution. Cato, Washington, D.C

Sowell T (2009) Applied economics. Basic Books, New York

The Declaration of Independence and the Constitution of the United States of America (2002) Cato, Washington, DC

Thuesen G, Fabrycky W (2001) Engineering economy, 9th edn. Prentice Hall, Upper Saddle River

Williams W (2000) The legitimate role of government in a free society. Imprimis 29(8):1–4

Wolmar C (2010) Blood, iron, and gold. Public Affairs, New York

Chapter 9
Effects of Taxation on Cash Flows

In this section, we will address the question of why we should include this topic in an engineering economy context. What uniquely Austrian considerations can be added to a more typical treatment of the problem?

9.1 Why Consider Taxes?

In a general sense, taxes are a cost to a project, or to our personal cash flow. Often, students have had little or no experience with how heavy the tax burden can become, or even how to compute their income tax bill. They need to know how high the tax rates can become, and I want them to be able to make an adequate evaluation of the effects of taxation in their economic calculations, whether prospective or retrospective.

Taxes distort incentives. The buyer must pay more, while the seller gets to keep less. Therefore, mutually beneficial, voluntary economic transactions that would increase the overall well-being of the economy would not occur. They become unprofitable and are a cost to the economy. Thus, price distortions are another component of the total cost of government (Gwartney et al. 2010).

We can use simple example calculations to demonstrate that taxes can affect the profitability of various projects. Taxation can divert investment of capital in several ways, one of which is to prevent it entirely for some projects. Desirable projects unknown to us according to Austrian economists would not be developed. No one can ever know what did not happen because of the reallocation of resources due to taxation. We do know that the project not developed would have been more highly valued than the project funded by tax revenue (Mises 1996).

The concepts we use to evaluate effects of taxes on engineering projects also apply to our personal tax reporting and financial decision making. For example, when we learn how to amortize a loan, we can imagine that it represents a home

© Springer International Publishing Switzerland 2015
R.A. Chadderton, *Purposeful Engineering Economics*,
DOI 10.1007/978-3-319-18848-5_9

mortgage. We might also suggest that the national debt (sic) should be considered to be a mortgage, and the amount of that debt is the present worth of discounted future taxes. We might also distinguish between the annual budget deficit and the total amount of the debt.

9.2 Depreciation Accounting for Capital Preservation

The necessity for economic calculation has been emphasized throughout this narrative as it has been in my class. Economic calculation is a fundamental insight of Austrian Economics. To be reliable, it must be based on market price data.

I once heard a conversation between a faculty friend and the parent of a student. In answering a question my friend replied, "I did not give your son an *F*, he earned it." I often used this story to introduce students to the necessity for a reliable economic calculation.

The fundamental idea is that profit and loss go together. The chance to earn a profit also means the risk of earning losses. The only way to determine whether you made a profit is with economic calculation based on your specific knowledge. The Austrians believed that ultimately an interventionist state would create economic chaos by making economic calculation impossible. Without prices such a society would operate in an information blackout, the result being "planned chaos" (Mises 1981).

For my purposes, the major causes of price distortion have been inflation and taxation. Specifically, inflationary effects can make a loss appear to be a profit. We can partially overcome these effects by accounting for allowable depreciation. Therefore, we need to have students understand depreciation as a tax-saving device, know how to calculate depreciation deductions, and be able to estimate resulting effects on taxes due.

We make the distinction between physical (wear and tear) depreciation and the methods of bookkeeping, or accounting, depreciation. Our objective is to recover the value lost or used in the work so that capital value is not consumed by taxes and inflation.

I emphasize that my class is not a tax course. To demonstrate the value of depreciation accounting, only straight-line (SL) and an accelerated method (MACRS) are compared. Students are asked to calculate annual depreciation allowance and book value tables for an equipment purchase with a salvage value. The questions they are asked to address are as follows:

What are the present values of the depreciation tables for SL and MACRS accounting methods?
What is the present value of taxes required to be paid, or what is the present value of the tax savings resulting from depreciation for both SL and MACRS methods?
What would happen if you were to sell the equipment for more than its book value at some time prior to the final year of the accounting period?

Students should see that although the total dollar amount of depreciation is the same for both methods, the present value of the depreciation table is higher for the accelerated method of MACRS because of basic time value. Their calculations should be sufficient to demonstrate the benefit of accelerated depreciation methods to reduce taxation and to minimize capital consumption due to inflation and taxes. Some of these effects are demonstrated in the examples below. SL depreciation is used for simplicity, while knowing that it is not the optimal approach.

9.3   Estimating the Effects of Taxation

Our major concern is to be able to convert a pre-tax cash flow into an after-tax cash flow. In addition to the basic interest factor equations, we will consider marginal tax rates, depreciation, interest payments, and inflation. We want an after-tax cash flow incorporating all these concerns to use in our decision rule to determine how best to proceed with our particular project.

We noted in Chap. 6, according to Hayek's (2009) molasses analogy, that inflation affects different components of a cash flow and different individuals to different degrees. All prices do not increase together at the same pace. These effects will cause distortions in the structure of price data, particularly disturbing exchange rates among goods from what they should be and consequently interfering with essential information transfers among actors in market processes. So, when we are faced with a taxed, inflationary environment (the "real" world), we want to use extra caution and emphasize that our predictions (prospective calculations) could have a wide uncertainty range.

Taxes are calculated on the actual, or current, money amounts. Therefore, an actual dollar analysis is required; any money amounts expressed in constant or base-year units (dollars) must first be converted to current, actual amounts before taxes can be calculated. Then, after accounting for anticipated taxes, the before-tax cash flow can be converted to an after-tax cash flow for evaluation. One could convert this actual dollar after-tax cash flow to a constant dollar after-tax cash flow for analysis, if desired. Of course, knowledge of a year-by-year rate of inflation would be required for this conversion.

When dealing with taxation in an inflationary environment, one must be explicit about what the acceptable minimum attractive rate of return, or prevailing interest rate, refers to: is it after-tax, actual, or constant dollar based? The result of moving from an imaginary world of no inflation or taxation into a real world with taxes and inflation is examined in the following example.

Example 9.1 We are considering a purchase of equipment with initial price of $20,000 and zero salvage value after a three-year life. We expect a constant dollar income of $9000 per year. The required minimum attractive rate of return (i') in the constant dollar domain is 8 % annually. Inflation is anticipated to average 5 % annually. Should we make this acquisition?

(a) Begin with the no tax, no inflation case:

"net P" $= \text{NPV}@8\% = -20,000 + 9000 * (P/A, 8\%, 3) = +\$3194$

Therefore, the acquisition is profitable.

(b) But, the expected marginal tax rate is estimated as 38 %. There is still no inflation.

$\text{NPV}@8\% = -20,000 + 5580 * (P/A, 8\%, 3) = -\$5619.$

So, now the investment would not be profitable.

(c) Depreciation of the equipment by SL method ($6667 per year is allowable) without any inflation gives:

$\text{NPV}@8\% = -20,000 + 8113 * (P/A, 8\%, 3) = +\$908.$

Now, the investment is profitable, again!

(d) Finally, we approach the "real world" of 5 % annual inflation and 38 % marginal taxation. The following table shows the sequence of calculations necessary to evaluate the proposed investment. The columns are as follows:
t (years); A (constant dollar amounts); A' (actual, inflated dollar amounts); D (depreciation); $A' - D$ (actual dollar taxable amounts); T' (tax); F' (actual dollar after-tax amounts); and F (deflated, constant dollar after-tax amounts).

t	A	A'	D	$A' - D$	T'	F'	F
0	−20,000	−20,000					−20,000
1	+9000	9450	6667	2783	1058	8392	7992
2	+9000	9923	6667	3256	1237	8686	7878
3	+9000	10,419	6667	3752	1426	8993	7768

Discounting F (constant dollar amounts) at 8 % annually by repeated application of Eq. 3.1 gives a $\text{NPV}@8\% = +\$320$. Would this investment be made?

We can point out that under inflationary conditions, both the investor (F') and the government (T') are receiving more paper dollars in comparison with case (c)! I ask the students why this is not good for everyone involved. However, when we investigate the situation using deflated (constant) dollars, we see that the effect of inflation has essentially eliminated the attractiveness of the investment by devaluing the anticipated future after-tax income. The "real" value has been virtually destroyed.

9.4 Summary of Effects of Inflation

Example 9.1 demonstrates several effects, as it is augmented stepwise from the imaginary, "no tax no inflation" world to the "real world."

(a) In the basic untaxed, non-inflationary example, the project specified exhibited a positive net present value, thus making it a worthwhile investment.

(b) When taxed at a marginal rate above 13.8 %, found from an internal rate of return calculation, the project (still without any inflation effect) would have a negative net present value making it a poor investment that would not be made.

(c) Incorporating SL depreciation of the equipment purchased (still without any inflation effect) made the acquisition worthwhile, even at 38 % tax rate, but its "profitability" shown by the net present value was substantially reduced from case (a).
(d) When modest inflation (5 % annual) was added, even with depreciation included, the net present value was reduced nearly to zero, compared to case (c).

We should remind the students that all of these example calculations are prospective and the realized net present value would only be known after the fact by retrospective calculation.

9.5 How Inflation and Taxes Affect Capital Values

Example 9.1 demonstrated that taxation and inflation could prevent investment. Taxation and inflation can also cause what I term "Capital Consumption" of past investments, as the following example shows.

Example 9.2 Equipment purchased for $60,000 is expected to have zero salvage value after a six-year life. Thus, SL depreciation of $10,000 per year would give a book value of $30,000 at the end of year 3.
Assuming equal loss of "utility" each year and the average inflation rate (5.91 %) from examples in Chap. 6, we project a sale price at the end of year 3 of:

$$\text{Sale Price} \sim 0.5(60{,}000) * (1.0591)^3 = \$35{,}640 \text{ (actual, current)}$$

The capital gain from inflation = $5640 is taxable at 34 % (Au and Au 1992)

$$\text{Tax} = -1918$$

Therefore, the actual value of sale = $33,722.
The capital consumed was approximately, $1918. The equipment could not be repurchased the next day because of this loss! The situation has been explained by Au and Au (1992):

"In a period of inflation, the sales price of an asset as expressed in then-current dollars increases but the book value is not allowed to be indexed to reflect the change of price level. Consequently, capital gains tax increases with the surge in sale price resulting from inflation."

9.6 Summary for Students

This section has demonstrated some effects of taxes and inflation on existing and proposed investments. We should conclude this discussion by reminding students that taxes are not the only cost of government. Gwartney et al. (2010) cite three costs of government:

1. The loss of private sector output lost to government supplied goods and services.
2. The costs of collection, enforcement, and compliance with government mandates.
3. The costs of price distortions that make some mutually advantageous exchanges unprofitable so that they do not occur.

References

Au T, Au T (1992) Engineering economics for capital investment analysis, 2nd edn. Prentice Hall, Englewood Cliffs

Gwartney J, Stroup R, Lee D, Ferrarini T (2010) Common sense economics. St. Martin's Press, New York (revised edn)

Hayek F (2009) A tiger by the tail. Ludwig von Mises Institute, Auburn

Mises L (1981) Socialism: an economic and sociological analysis. Liberty Fund, Indianapolis

Mises L (1996) Human action, 4th edn. The Foundation for Economic Education, Irvington-on-Hudson (revised edn)

Chapter 10
An Example "Retirement" Planning Calculation

Much current advertising asks "How much savings will you need for your retirement?" One personal benefit for our students from this course should be to learn how to estimate such a future need and to calculate a cash flow to reach that goal. It should be obvious that if savings grow 4 % while the CPI increases 5 %, value has been lost.

10.1 Account for Devaluation of Money

To begin this lesson, I ask the students what the symbol $ means. They usually say "dollars," while the origin was probably the abbreviation for USA. The next question is to ask "What is a dollar?" A dictionary will probably say something similar to "a coin or note worth one dollar" (American Heritage 1985), or a "dollar equals 100 cents." This is not the kind of answer desired. Originally the dollar was defined as one-twentieth ounce of gold. Thus, the original price of gold was $20 per ounce. This definition process was not unusual as currencies were typically defined in terms of weights of precious metals. Of course, today the price of gold is much higher. What happened? In terms of simple supply and demand, the answer would be that either there is a shortage of gold or a surplus of dollars.

We could, and should, call the process a <u>devaluation</u> of the dollar as justifiably as saying that the price of gold has skyrocketed. We should reiterate that Gresham's Law tells us why there are no gold (or, silver) coins in circulation today. An appropriate restatement of Gresham's Law (Greaves 1974) for this purpose would be the following: An <u>undervalued</u> resource will not be used it will be hoarded (e.g., gold or silver coins). An <u>over-valued</u> resource will be used as a substitute (e.g., copper or nickel coins, or paper dollars).

© Springer International Publishing Switzerland 2015
R.A. Chadderton, *Purposeful Engineering Economics*,
DOI 10.1007/978-3-319-18848-5_10

We should attempt to incorporate the effect of devaluation in our retirement account planning. The following examples are intended to demonstrate the effect of modest inflation rates on "retirement income."

Example 10.1 Calculate the actual monthly after-tax income needed to maintain a base-year value of $1000 units for various annual inflation rates (f).

Time (years from now)	$f = 3\%$	$f = 4\%$	$f = 5\%$
0 (now)	$1000	$1000	$1000
5	$1159	$1217	$1276
10	$1343	$1481	$1628
15	$1557	$1802	$2078
20	$1804	$2194	$2651
25	$2091	$2670	$3383

Note: We can define a 5-year inflation rate for this example as $(1 + f)^5$, thus:

$$(1.03)^5 = 1.159$$
$$(1.04)^5 = 1.217$$
$$(1.05)^5 = 1.276$$

Example 10.2 Calculate the base-year (constant dollar) value of a "retirement income" of $1000 per month (after tax) for various annual inflation rates (f).

Time (years from now)	$f = 3\%$	$f = 4\%$	$f = 5\%$
0 (now)	$1000	$1000	$1000
5	$863	$822	$784
10	$744	$676	$614
15	$642	$555	$481
20	$554	$456	$377
25	$478	$375	$295

Note: We can define a 5-year devaluation rate for this example as $(1 + f)^{-5}$, thus:

$$(1.03)^{-5} = 0.8626$$
$$(1.04)^{-5} = 0.8219$$
$$(1.05)^{-5} = 0.7835$$

The previous two examples were not intended to cause fear but rather to raise awareness of the implications of even modest rates of devaluation. Clearly, if one can anticipate the rate of inflation and is able to increase income to overcome the devaluation, the constant dollar value of $1000 could be maintained. We have assumed that an amount of savings adequate to provide the necessary annual withdrawals is available at time 0. Methods to calculate the cash flow needed to generate that total savings amount were shown in Chap. 5 as an equivalence calculation. A calculation of the annual account balance during the withdrawal period was also shown in Chap. 5.

10.2 Account for Taxation

Example calculations have not yet accounted for the effect of taxation anticipated in the future. Therefore, the next step will be to consider the compound effect of inflation (devaluation) and taxation on investment earnings. For convenience we can use the CPI data for the period 1968–1992, again (Thuesen and Fabrycky 2001). We showed previously (Chap. 6) that the average inflation rate based on past price increases was 5.91 % annually for this 25-year period. We must remind students that any historic, average rate is still only a guess when we make prospective calculations.

Example 10.3 demonstrates that a 5 % investment income taxed at 25 % marginal rate actually lost value at a rate of 2.16 % annually. The loss would be even greater if we were fortunate enough to be in the 28 % marginal tax bracket. Students should make that enlightening calculation for themselves. It should become apparent that we must attempt to anticipate to the best of our ability both taxes and inflation when we make projected income requirement calculations.

Example 10.3 Estimate a rate of return for a projected investment of $1000, today.

Assumptions : average $f = 5.91$ % annually (an historic average)

Marginal tax rate $= 25$ %

Average rate of return ~ 1.1 % per quarter per today's information

Annualized $= (1.011)^4 = 4.47$ % annually (use 5 %)

Income @ 5 % return $= \$50.00$

Tax @ 25 %, federal $= \$12.50$

Net, after tax $= \$37.50$ (actual dollars)

Devaluation @ 5.91 % $= \$59.10$

Net, after tax income $= -\$21.60$ (constant, base year dollars)

Thus, the real rate of return is -2.16 % annually. Real value was lost.

10.3 Estimate Combined Effect of Devaluation and Taxation

Finally, we should consider taxes and inflation together and ask how much return we would need to maintain a constant value retirement income of $1000 per month? What pre-tax actual amount ($X) one year later will be needed to provide a base-year value of $1000 (after tax)? The results of calculations shown below are $1412 for a 25 % tax rate and $1471 for a 28 % tax rate.

$1000 today (after tax) $\rightarrow$ $1059.10 (after tax, one year later)

Pre-tax $= \$1059.10/0.75 = \$1412.33$ (one year later)

<u>Note</u>: At 28 % tax rate; required pre-tax = \$1470.97 (one year later).

Does this required income actually preserve current value? We can verify the calculation as follows:

$$\text{"Pre-tax income"} = \$1471/\text{month (one year from now)}$$
$$\text{Tax at 28 \%} = \$412$$
$$\text{"After-tax"} = \$1059/\text{month}$$
$$\text{"Deflated"}@5.91\ \% = \$1000/\text{month (checks)}$$

10.4 Concluding Remarks

Possibly the previous calculations should use an average tax rate, such as 15 %, that would be realistically lower than the marginal rates actually used. However, the main point would be the same; one must overcome both taxes and inflation to maintain constant value over time. The number of actual dollars that would be needed in the future will grow as we extend the time line. For example,

$$F\ (t \text{ years later}) = \$1000(1.0591)^t/(0.75)$$

e.g.,

$$F\ (10 \text{ years later}) = \$1000(1.7757)/(0.75) = \$2368 \text{ at 25 \% tax rate}$$

or

$$F\ (10 \text{ years later}) = \$2466 \text{ at 28 \% tax rate.}$$

Finally, we should remind students that the present value for these calculations is actually required at the beginning of the first year that retirement savings are withdrawn.

References

American Heritage Dictionary, Second College Edition (1985) Houghton Mifflin Company, Boston

Greaves P (1974) Mises made easier. Free Market Books, Dobbs Ferry

Thuesen G, Fabrycky W (2001) Engineering economy, 9th edn. Prentice Hall, Upper Saddle River

Chapter 11
Costs of Regulation

When engineers work in fields regulated by the general, federal government, they favor regulations for scientific or technical reasons. Since regulations are frequently derived from risk analyses, it would be valuable to society if engineers understood the total effect of regulatory costs on the general populace before lobbying for more regulation and public expenditures. The purpose of this section is to suggest within the context of engineering economics a broader perspective on the costs of regulation.

Gwartney et al. (2010) identified three types of costs caused by governments' taxation: the loss of private sector output; the costs of tax collection, enforcement, and compliance; and the costs of price distortions. As a result, they note that in "2004, the cost of enforcement and compliance with tax laws and regulatory mandates was estimated to be....11 % of national income." (p. 105)

While regulations are considered to be necessary for some so-called non-market, public goods, they interfere with free markets, often having adverse side effects. This is another example of the "seen" versus "unseen" consequences as described by Bastiat (2011) and Hazlitt (1996) introduced in Chap. 3.

11.1 Risk Evaluations and Perceptions

This section is not intended to be an extensive examination of risk decisions. However, because many regulations and their attendant costs derive from risk calculations, some descriptive material is necessary prior to considering the costs of regulations.

Wildavsky (1995) developed numerous case studies and concluded that expending limited resources on tiny "risks" actually makes the world less safe, causing so-called "statistical deaths". The risk portion of my course includes

© Springer International Publishing Switzerland 2015
R.A. Chadderton, *Purposeful Engineering Economics*,
DOI 10.1007/978-3-319-18848-5_11

discussions of the relationships among fear, risk, and safety according to Wildavsky. Expensive reduction of small risks mandated by overly restrictive regulations can result in increased death from greater risk factors. An excellent summary of these arguments can be found in an article by Smith (1995).

For several years, I used a film called "Are We Scaring Ourselves to Death" in class. This documentary investigated several risks that have been promoted in media as being much more dangerous than assessments by risk specialists indicated. An article relating the evolution of the narrator's attitude toward risk and the conclusions drawn in the documentary was used as a supplemental reading in the class (Stossel 2001).

Fear is a limited resource. Each person has a prioritized list of fears; to add a new fear, others must be moved down in the ranking. One cannot fear everything. Consequently, too many exaggerated risk scares can cause fear of innovation. For example, strict application of the Precautionary Principle could mean that old technologies, such as gas heating or automobiles, probably would not be acceptable today. Possible other harmful effects of the Precautionary Principle could include restrictions on new vaccines and antibiotics (Beckerman 2003).

Bad things happen, but too much fear is a bad thing. Too much regulation is also a bad thing but too few people fear that. Expenditures contribute to price distortion but the costs are more than money, they include statistical deaths. Consequently, engineers should appreciate that the regulation of risk is not risk free.

Wildavsky's analysis of the jogger's dilemma can be used to introduce students to the study of risk. Jogging is dangerous in the short term but gives long-term benefits. The dilemma shows the interconnection of risk and safety. Society's safety has been achieved by applications of risky technology requiring vast expenditures of limited resources. Wildavsky's (1988) conclusion that "wealthier is healthier" can be used to describe the interaction of economics and risk. Because resources are always in short supply (Mises 1996), as is fear itself (DeBecker 1997), the conclusion should be that fearing and expending resources on low-risk hazards could actually reduce overall safety (Chadderton 2003).

Students can be introduced to the concept of "statistical deaths" and the possibility that expensive reduction of small risks mandated by overly restrictive regulations can result in increased death from greater risk factors. Public perceptions of risk are often unrelated to actual risk (Slovic et al. 1979). Fischhoff et al. (1981) show that being an unmarried male results in 3500 days of "lost life expectancy" while alcohol use could cost 130 days, on average. Smoking could cost 2250 days and cancer 980 days, while "low socioeconomic status" could cost 700 days (Fischhoff et al. 1981).

Expending all resources to combat costly, low-risk hazards would leave insufficient resources to address mundane, yet substantial, risks. Flexibility and resilience are lost, leaving no ability to respond to unanticipated risks (Wildavsky 1988). Expensive programs to reduce small, dispersed hazards will actually make people poorer and less healthy. (Chadderton 2003) Wildavsky ultimately concluded that the attempt to reduce risk to zero was impossible to achieve and that zero risk was the greatest risk of all (Wildavsky 1990).

Additional perspective on risk can be gained from Stossel (2001). For example, toxic waste sites might cause a loss of four days from an average life, house fires might claim 18 days from the average life, but driving could be credited with the loss of 182 days of the average life. Based on these numbers, we could suggest that to reduce risk and improve life expectancy, we need to outlaw "being unmarried".

When the public's rankings of perceived risks are compared to experts' risk assessments, the rankings are practically reversed. For example, college students ranked nuclear power as a much higher risk than experts did. The experts ranked motor vehicles, alcohol, and swimming as higher risks than did the students (Slovic 1987).

The purpose of this selection of perceived versus actual risks is not to give a complete analysis but rather to suggest the divergence from activities seen as high risk to those that really are high risk. This disconnect could lead, via political activity, to the expenditure of scarce resources on the regulation of low hazard "risks".

An alternative way to consider risk is the estimated "cost to save a life". Tengs et al. (1995) compiled cost-effectiveness of 500 lifesaving interventions. For example, many disease screening and health improving measures were estimated to be orders of magnitude more cost-effective than many emission control measures (Tengs et al. 1995). Wildavsky (1988) also included estimated costs to save lives. For example, airbags were estimated to cost ten times more than cardiac emergency units and 30 times more than cancer screening programs to save a life (Wildavsky 1988).

One conclusion that can be drawn from these numbers, and should be emphasized with students, is that mundane, local actions can save lives at less cost than can exotic, global actions. We should also note that as the scale (size) of a problem increases, the uncertainty bound around its anticipated cost also increases (Chadderton 1993a). This seems to indicate that attempts to solve global-scale problems have a greater chance of failure than attempts to solve local problems.

Ratios of lost life expectancies can be calculated to emphasize that the regulation of risk has an exchange value in terms of lives saved. For example, the exchange value of an air bag for a cardiac emergency unit would be about ten lives lost. This approach assumes that saving lives efficiently is the objective.

11.2 Wage Control and Lifespan

We note that one of the greater causes of reduced lifespan is poverty (low socioeconomic status, above). Another form of government regulation that has unseen consequences on socioeconomic status, namely minimum wage controls, was considered in Chap. 7. This is an often repeated fallacy that deserves to be revisited here. Politicians continue to promise higher wages by raising the minimum allowable wage rates. Many economists have exposed the consequences for countless unskilled workers, especially among minority groups.

Sennholz (1995) described the situation. "Few economic laws, if any, are more malicious and malignant than minimum wage laws. They prohibit workers from accepting employment unless they are paid at least the minimum. They order employers to use only workers who qualify for the minimum and reject all others. The laws erect a hurdle over which all American workers are forced to jump." Potential workers who cannot justify the total cost of the minimum wage (total cost, including all mandated "benefits," not just salary) will not be employed. We should point out to students that an employer must justify the total cost of the marginal employees (Sowell 2000). The result of extremely high rates of unemployment, particularly among minority youth, is idleness and often crime (Sennholz 1995; Sowell 2000; Williams 2000). The result among the general populace is taxation to support the institutionally unemployed and fear of crime.

Simple supply and demand considerations show that a higher promised salary will attract additional less qualified workers while reducing the number of available positions. Labor unions support higher minimum wage rates which enhance the power of unions to control entry into unionized labor markets (Sennholz 1995; Sowell 2000; Williams 2000). However, unionization cannot raise the wages of all willing workers. Those not employed in unionized markets are shifted into non-union jobs and force unregulated wages downward.

The additional effect of the minimum wage cycle, institutionalized unemployment as shown above, is a lowering of socioeconomic status and resulting "lost life expectancy." This should remind us of Bastiat's (2011) warning to look for the "unseen." We should urge students to consider the potentially harmful, long-term effects of even well-intended short-term policies.

11.3 Examples of Regulatory Effects

Engineers should appreciate all of the effects of government interventions of various kinds in the economy. (Chadderton 1994) One purpose of my course has been to offer a few examples of the unanticipated, long-term consequences of government funding and regulation of apparently beneficial programs. The following sections introduce several examples of regulatory actions and suggest alternatives that are frequently discounted by advocates of stronger regulation policies. The examples have been abbreviated, here. References to more complete presentations are provided.

11.3.1 Transcontinental Railroad Subsidies

Governments have been involved with railroad developments in many countries. While numerous reasons were used to justify these interventions, a typical argument was that "the state had to be involved" in such important infrastructure (Wolmar 2010).

Some of the specific reasons for government involvement were that railroads could be a political tool to unify a nation (Germany, Italy), they could expedite moving military forces (internally or for national defense), and they would promote economic interdependence (the German states). Lack of capital was considered to be a common problem that necessitated government involvement (e.g. Spain, Russia, Scandinavia).

Typical results of these government supported projects were hurried, poor construction, premature overbuilding, corruption, bankruptcies, and consequently regulation by the government. These results have been universal, not unique to the United States (Holbrook 1947; Martin 1992; Wolmar 2010).

Governments on the continent of Europe were typically more involved than in Britain where the government took a more disinterested and laissez-faire attitude toward railroads. The references contain much historical information for a number of countries. The concern for us is to try to discover what Austrian economists would conclude about the interventions of governments in railroad development.

This case concerns the transcontinental systems in the United States. Mercer (1982) estimated the costs of federal regulation for the American transcontinental systems. In addition to economic efficiency, effects of land grants on income distribution, land settlement patterns, and constitutionality of the subsidies and ethics of an interventionist policy are major concerns in this case.

A comparison of the Northern Pacific and the Great Northern systems can be used to demonstrate opposing financial policies for transcontinental railroads in the nineteenth century. The two systems were essentially parallel and not far apart. The Northern Pacific was highly subsidized, while the Great Northern was created with minimal subsidies (possibly none). Details of the construction and subsequent history of the two systems can be found in various sources (Mercer 1982; Brown 2001; Holbrook 1947; Folsom 1987; Martin 1992; Rand 1967).

The transcontinental systems were caught up in the tumultuous period following the Civil War. In short, the railroads with the worst histories of scandal and bankruptcy were those that had received the greatest amount of help from the government.

The United States Congress had financed the Civil War by borrowing money, by issuing treasury notes (greenbacks), and by creating a national banking system that was able to monetize the war debt by issuing bank notes backed by government bonds. Carson (1971) maintained that these methods of financing were inflationary, unconstitutional, and a major extension of the power of the federal government. The results were short-term prosperity and expansion followed by panic and depression. This sequence of inflation and panic can be understood as a specific case of "boom and bust" and is best explained by the Austrian theory of the trade cycle (Taylor 1980; Mises 1996).

Mises (1996) maintained that "Public works are …paid for by funds taken away from the citizens. If the government had not interfered, the citizens would have employed them for the realization of profit-promising projects….For every unprofitable project that is realized by the aid of the government there is a corresponding project …which is neglected merely on account of the government's intervention." Thus, the public project is seen, while the neglected project is unseen. No one can know what profitable projects never materialized because of the intervention.

In 1860, a majority of voters favored subsidies for the railroads. But after the Panic of the 1870s, railroads were hated because they were subsidized. That hatred lasted until the railroads were regulated and controlled by the Interstate Commerce Commission (Lane 1984). Ironically, since deregulation in the 1980s, the railroads have become innovative and efficient. The public has benefited from renewed entrepreneurship (Martin 1992).

The subsidized, politicized railroads went bankrupt in the Panic of 1893, while the Great Northern was financially sound and did not. One can determine from an Austrian assessment that without subsidies, the transcontinental railroads would have been built later with better quality materials and workmanship. White (2011) concluded that the transcontinental railroads would not have been needed until much later. Then, they could have been built more cheaply, more efficiently, and with fewer social and political costs.

These conclusions are not restricted to the railroads, however. Carson (1971) summarized the effects of ill-conceived government intervention on taxpayers: "though canals and railroads often did get built, the government involvement not only spurred hasty and ill-considered building but also placed heavy burdens of taxation on their citizens."

An appreciation by engineers of the unanticipated, adverse effects of government interventions on the economy would ultimately further the best interests of society. These issues should be recognized in engineering education.

Additional detail about the transcontinental railroad case study and the supporting references can be found in Chadderton (1993b).

11.3.2 Water Pollution Control

Water pollution has been considered a prime example of market failure; the costs of externalities being borne by others than the polluter. Typical solutions to pollution have been command-and-control schemes involving new bureaucracies created to administer these programs, essentially forever. In this case, we ask whether an economically justifiable alternative method could exist. A hypothetical, although realistic, water pollution problem is studied from a property rights viewpoint.

Almost any environmental engineering textbook will include a description of and a justification for the regulatory process. The regulatory program in a stream pollution case typically derives from some sort of waste load allocation (WLA) process. Each WLA method would lead to different regulations for dischargers. Studies of alternative WLA methods attempted to find the most cost-effective WLA procedure (Chadderton et al. 1981; Chadderton and Kropp 1985).

The main justifications for pollution controls have been preservation of resources, aesthetics, and health. However, further study has documented evidence that the primary value of instream flows is from recreation and fishing; therefore, economic justification for controlling stream pollution depends on recreational benefits (Kneese and Bower 1968; James and Lee 1971). Further, the evaluation of recreational benefits is problematic at best and virtually impossible in practice (Goodman 1984; Kneese and Bower 1968). Dissatisfaction with results of the command-and-control regulatory process caused concerns that methods and institutions for water quality management needed to be changed (Viessman and Hammer 1985; Viessman and Welty 1985).

A hypothetical stream pollution problem and the accompanying data come from a "model of public decisions" in a water pollution policy problem by Dorfman and Jacoby (1972). The problem was simplified enough to be mathematically solvable but detailed enough to present elements of conflict among competing claimants to the river resource.

All data and numerical assessments included in Dorfman and Jacoby (1972) were used without question. Assumed low flow in the river, results of water quality modeling of the biochemical-dissolved oxygen system (BOD-DO), treatment cost schedules for three dischargers, and tabulated recreational benefits as a function of instream water quality were used unchanged.

This analysis was intended to evaluate a property rights approach to the stream pollution problem. An environmental or recreational group owner was postulated initially. Fully liable private owners were assumed both upstream and downstream from the study area. Acquisition and transfer of property rights to water was assumed to be permitted.

An estimate of maximum recreational benefits was calculated from a benefits table provided by Dorfman and Jacoby (1972). This showed that a recreational group could not afford to provide the required treatment facilities needed to maintain acceptable DO levels in the stream. They could be outbid for the resource by the dischargers who could thereby avoid treatment.

Ownership of the river segment was transferred to a water supply company that would be fully liable to owners of adjoining stream segments. Based on typical water usage and price of water, the water supply company would generate sufficient income to provide adequate treatment at all discharge sites, maintain instream DO levels, and generate recreational facilities. The water company could outbid the dischargers and achieve control of the stream segment.

Of course, it is impossible to predict exactly what would happen under private rights because of fundamental uncertainty in future human action. However, the Austrian economists have maintained that bureaucratic planners suffer the same limitations to prediction. In the longer term, market actions would discover the most urgent, economic use of the resource.

In this particular case, a fully liable, private owner was found to be financially profitable while improving the water quality and providing recreational potential. The example confirmed that property rights provide an answer to stream pollution problems, but contradicted the view that water pollution control must be justified by recreational benefits. Additional analysis of this case can be found in Chadderton (1988).

More recent references on pollution control are considered here. For example, Gwartney et al. (2010) emphasize marginal decision-making even in pollution control cases. "...as the amount of pollution goes down, so does the marginal benefit—the value of additional improvement..." And, "...while the marginal benefit of reducing pollution is going down, the marginal cost is going up and becomes very high before all pollution is eliminated."

If the marginal cost exceeds the marginal benefit, the incremental improvement (reduction of pollution) is not desirable. There would be too much regulatory cost. A series of example problems from Rubin (2001) has been used in my class to demonstrate these effects.

A table of "Median Cost/Life-Years Extended" (Stroup 2003) shows that the cost to save an additional life-year was $23,000 for FAA regulations and $88,000 for OSHA rules to reduce fatal accidents. However, EPA regulations showed an estimated cost of $7,600,000 for each additional life-year extended. Stroup (2003) concluded that more life-years could be saved by a reallocation of regulatory costs or that the same number of life-years could be saved at significantly reduced costs by shifting resources. This is the same conclusion stressed by Wildavsky (1990) that administrative regulation is causing many Americans to die prematurely because excessive amounts of resources are expended to achieve small marginal gains in risk reduction.

Stroup (2003) also notes that a major benefit of market solutions to environmental problems is that "market decisions are diverse and decentralized." They offer a variety of approaches instead of a bureaucratic large, uniform approach that can cause great harm if wrong. Stroup (2003) concluded that "although property rights and markets are imperfect, we have seen evidence that they are greatly underappreciated, and thus greatly underused at this point in the history of environmental policy."

In referring to pollution problems, O'Driscoll and Rizzo (1996) argued that "absence of markets and improperly specified property rights do generate economic problems. The very absence of relevant markets, however, implies the absence of any ability to acquire the very information needed to correct the problem. If markets are not providing signals to transactors on all costs of an action, policy-makers will also lack this information. Moreover, even were they to possess the required information, political decision-makers face different incentives than do entrepreneurs."

These incentive effects have been shown in an analysis of the EPA Superfund program by Stroup (1995). I call them perverse incentives (Chap. 8) that should be explained to our engineering students.

11.3.3   Land-Use Regulations

The basis for another interesting example concerning land-use regulations was recently published (Turner 2014). The author reported on research that had assembled and analyzed massive amounts of information about land values and land-use regulations from various national databases. Both the value of land and the level of regulation on land-use were considered to be very high in the United States.

An assessment of three effects of regulation on land values (namely what was termed the "own-lot effect," the "external effect," and the "supply effect") led Turner (2014) to state that overall the own-lot and external effects are negative. He concluded that less regulation of land use might be better. The benefits from modest reductions in land-use regulations are likely to be greater than the benefits of continuing, stronger regulation (Turner 2014).

These conclusions by economists are likely to be counterintuitive to engineering students who generally seem to have a bias in favor of regulations. Therefore, this is another example that could be presented to our engineering students to show that over-zealous regulators could actually cause negative effects on the economy.

References

Bastiat F (2011) That which is seen, and that which is not seen. Dodo Press, UK

Beckerman W (2003) A poverty of reason. The Independent Institute, Oakland

Brown D (2001) Hear that lonesome whistle blow. Henry Holt, New York

Carson C (1971) Throttling the railroads. Liberty Fund, Indianapolis

Chadderton R, Miller A, McDonnell A (1981) Analysis of waste load allocation procedures. Water Resour Bull 17(5):760–766

Chadderton R (1983) Praxeology and engineering. J Prof Issues Egr 109(3):159–169

Chadderton R, Kropp I (1985) An evaluation of eight wasteload allocation methods. Water Resour Bull 21(5):833–839

Chadderton R (1988) An alternative to water pollution controls. Water Resour Bull 24(1):183–187

Chadderton R (1993a) Misesian assessment of systems analysis. J Prof Issues Eng Edu Pract 119(4):346–357

Chadderton R (1993b) Land-grant policy and engineering education. J Prof Issues Eng Edu Pract 119(4):427–436

Chadderton R (1994) Relation of praxeology to engineering. J Prof Issues Eng Edu Pract 120(4):384–391

Chadderton R (2003) Should CEs study economics and risk? J Prof Issues Eng Edu Pract 129(4):198–202

DeBecker G (1997) The gift of fear. Little, Brown, and Co., Boston

Dorfman R, Jacoby H (1972) A public-decision model applied to a local pollution problem. In: Dorfman R, Dorfman N (eds) Economics of the environment, selected readings. Norton, New York

Fischhoff B, Lichtenstein S, Slovic P, Derby S, Keeney R (1981) Acceptable risk. Cambridge University Press, Cambridge

Folsom B (1987) Entrepreneurs versus the state: a new look at the rise of big business in America, 1840–1920. Young America Foundation, Reston

Goodman A (1984) Principles of water resources planning. Prentice-Hall, Englewood Cliffs

Gwartney J, Stroup R, Lee D, Ferrarini T (2010) Common sense economics, revised ed. St. Martin's Press, New York

Hazlitt H (1996) Economics in one lesson. Laissez Faire Books, San Francisco

Holbrook S (1947) The story of American railroads. Crown Publishers, New York

James L, Lee R (1971) Economics of water resources planning. McGraw-Hill, New York

Kneese A, Bower B (1968) Managing water quality: economics, technology, institutions. Resources for the Future, Washington DC

Lane R (1984) The discovery of freedom. Laissez Faire, New York

Martin A (1992) Railroads triumphant. Oxford University Press, New York

Mercer L (1982) Railroads and land grant policy: a study in government intervention. Academic Press, New York

Mises L (1996) Human action, 4th rev ed. The Foundation for Economic Education, Irvington-on-Hudson

O'Driscoll G, Rizzo M (1996) The economics of time and ignorance. Routledge, New York

Rand A (1967) Capitalism: the unknown ideal. Signet, New York

Rubin E (2001) Introduction to engineering and the environment. McGraw-Hill, New York

Sennholz H (1995) Minimum wages. In: Notes from FEE. The Foundation for Economic Education, Irvington-on-Hudson

Slovic P, Fischhoff B, Lichtenstein S (1979) Rating the risks. Environment 21(3):14–20, 36–39

Slovic P (1987) Perception of risk. Science 236:280–285

Smith F (1995) Risks in the modern world: what prospects for rationality? The Freeman 45(3):140–144

Sowell T (2000) Basic economics. Basic Books, New York

Stossel J (2001) The real cost of regulation. Imprimis 30(5):1–5

Stroup R (1995) Controlling risk: regulation or rights? The Freeman 45(3):149–151

Stroup R (2003) Eco-nomics. Cato, Washington DC

Taylor T (1980) The fundamentals of Austrian economics. Cato, San Francisco

Tengs T, Adams M, Pliskin J et al (1995) Five hundred life-saving interventions and their cost-effectiveness. Risk Anal 15(3):369–383

Turner M (2014) The economics of land-use regulations. PERC Reports 33(2):16–19

Viessman W, Hammer M (1985) Water supply and pollution control, 4th edn. Harper & Row, New York

Viessman W, Welty C (1985) Water management: technology and institutions. Harper & Row, New York

White R (2011) Railroaded. Norton, New York

Wildavsky A (1988) Searching for safety. Transaction Publishers, New Brunswick

Wildavsky A (1990) No risk is the highest risk of all. In: Glickman T, Gough M (eds) Readings in risk. Resources for the Future, Washington DC, pp 120–127

Wildavsky A (1995) But is it true? Harvard University Press, Cambridge

Williams W (2000) The legitimate role of government in a free society. Imprimis 29(8):1–4

Wolmar C (2010) Blood, iron, and gold. Public Affairs, New York

Chapter 12
Sustainability Concerns

Why include sustainability concerns in this book about engineering economy? We are not training sustainability specialists. I believe the case can be made that engineers, especially civil engineers, have always been implicitly sustainable; they do not design unsustainable projects intentionally. The primary reason to discuss sustainability might be the mandate by the American Society of Civil Engineers (ASCE) that engineering programs must include sustainability within an undergraduate curriculum to gain full accreditation. Another reason might be to ensure that engineering students have some idea about the multi-dimensional characteristic of "sustainability" as well as its limitations.

12.1 Defining Sustainability

Among the many definitions of "sustainability" are those of various engineering societies included in the ASCE Committee on Sustainability publication, *Sustainable Engineering Practice* (2004).

"Sustainable development... meets the needs of the present without compromising the ability of future generations to meet their own needs." (World Commission on Environment and Development (WCED) 1987).

"Development that will meet the long-term needs of future generations of all nations without causing modification to the Earth's ecosystems." (International Federation of Consulting Engineers (FIDIC)-1990).

"Sustainability is the ability to maintain a high quality of life for all people, both now and in the future, while ensuring the maintenance of the ecological processes on which life depends and the continued availability of the natural resources needed." (Institute of Engineers, Australia (IEA)-1994).

© Springer International Publishing Switzerland 2015
R.A. Chadderton, *Purposeful Engineering Economics*,
DOI 10.1007/978-3-319-18848-5_12

As noted by ASCE (2004), these definitions focus on environmental sustainability, while the additional dimensions of economical, political, social, and ethical sustainability should also be included (Hatch 2002). Another aspect of the sustainability question is the planning horizon; should it be 10, 100 years, or longer? The time period and the discount rate assumed would interact to have a major influence on present worth estimates.

12.2 What Is the Objective?

Questions related to sustainability that engineers should be willing to ask include "What needs", "What resources", and "When". For example, the idea of a triple bottom line calculation (see, for example, Davidson et al. 2007) introduces a concatenation of economic, environmental, and social concerns including the following:

- Economic: effective investments (engineering economics), essential finance, job creation, and competitiveness.
- Environmental: natural systems and public health (reduce use of non-renewable resources; better manage use of renewable resources).
- Social: equity, justice, security, employment, and participation.

These objectives could be conflicting. We must concern ourselves with the proper functioning of prices in the attempt to achieve sustainability goals.

12.3 A Question of Prices and Uncertainty

In Chap. 1, we considered the question of whether we could have certainty or only some degree of knowledge. The principle of tolerance indicated that we are forced to function within some range of uncertainty (Bronowski 1973). In Chap. 11, we introduced Wildavsky's (1988, 1990) discussion of anticipation versus resilience and concluded that inappropriate applications of the Precautionary Principle could lead to excessive emphasis and investment to prevent relatively minor risks. This mal-investment would leave inadequate resources to react to unanticipated, possibly greater risks that developed later.

We should argue that to be sustainable, we must prevent mal-investment. How can mal-investment be prevented? The simple answer is to allow the price system to work freely. Reviewing the functions of prices, especially for rising prices, will help to answer that question and will show the relevance of economic calculation for sustainability concerns.

12.4 A Dissenting View

In Chap. 7, we cited Hayek's (1977) analysis that to avoid confusion and mal-investment, price signals must not be distorted. The purpose of rising prices is to signal a shortage. Rising prices will stimulate the following events:

- Conservation by users,
- Exploration by producers,
- Substitution by innovators, and
- Hoarding by speculators.

All of these actions tend toward the creation of a new intersection of supply and demand by increasing supply and reducing demand. Simon (1981) called human ingenuity the "ultimate resource" essentially because of these responses.

Conversely, falling real prices signal an adequate supply or a pending surplus. This can explain increased demand for gasoline even at elevated nominal prices, an example calculation that my students have been asked to examine. Recently, even nominal prices have been falling apparently as a result of the combined effects of increased supplies and diminished demands. One would expect that these effects have been influenced by previously increasing nominal prices.

We have relied on Hayek's analyses of the distributed nature of knowledge and the values of the price system to coordinate the uses of that knowledge (1974, 1977). Prices distorted by interventions cause confusion and mal-investment; therefore, prices must not be distorted. I believe that engineering students need to be aware of these consequences of various types of intervention by government agencies.

I have used the Ehrlich–Simon 'bet' in class as another example to demonstrate the value of price functions. Essentially, the bet was over whether the real price of a selection of natural resources would rise or fall during a specified time period. Simon won the bet because the real prices of the resources fell during the time period. A recent book by Sabin presents a more complicated picture of this example. Sabin concluded that Simon was too optimistic and Ehrlich was too pessimistic. The resources and the time period selected for the bet were particularly favorable to Simon's argument (Sabin 2013). Sabin (2013) also concluded that these two opposing opinions continue to interfere with reasoned discussion of environmental issues, especially with respect to the climate change debate. However, he supported the position that excessive pessimism has a real cost (Sabin 2013).

The fear of running out of natural resources is nothing new; it is probably as old as civilization, itself. Mises (1996) wrote about excessive fear of resource depletion decades before environmentalism took center stage, noting that only those natural resources that can be used economically will be exploited, leaving unused reserves until the economics changed. Wildavsky (1995) and Stroup (2003) also support market mechanisms for resource utilization.

Solow (1992, 1993) maintained that sustainability is an essentially vague concept. He argued that it is not specific physical resources that must be preserved, but rather the capacity to be well off. Because of substitutability, this would include making replacements available (for example, technologies or knowledge). He suggested that investment of the rents from non-renewable resources in knowledge generation would be as environmentally beneficial as possible. Beckerman (2003) made similar observations.

We cannot know what future generations will want or need. This suggests an imperative to preserve a capability to satisfy unspecified future needs. It does not demand the sacrifice of current needs; future generations might be more well off than the current population (Beckerman 2003).

The treatise by Mises (1996) is too difficult to assign for students to study. However, the article by Solow (1993) is not. I have had the students in my classes read this article that explains his economist's perspective. Results on an exam essay question and responses to a brief survey showed that the students thought that it was a valuable assignment. It also broadened their perspectives on the issues surrounding the sustainability concept.

12.5 Differing Perspectives: Environmental Science Vs. Civil Engineering Students

A two-year assessment was made to determine whether engineering students modify their perspectives on sustainability because of exposure to topics of real economics and risk in their course work. Comparison of word association studies reported for environmental science students with the results from CEE students revealed interesting differences.

The environmental science department surveys (Sherman 2008) reported that 90 % of students associated the sustainability concept with prescribed practices such as recycling or composting. Only 5 % responded with a bigger idea, such as systems or balance and 5 % picked specific environmental problems such as global warming or biodiversity. Only 14 % of the CEE students identified sustainability with prescribed practices; 18 % chose environmental problems using words such as energy or green; while 56 % of the responses were long term or similar words such as renewable, durability, or reliability. The remaining 12 % were either non-responsive (8 %) or admitted having no idea (4 %). These engineering students generally had a much different view of sustainability than did the environmental science students (Chadderton 2010).

Because of their preparation in studies of real economics, and engineering economics, the engineering students had a larger idea of the problem of sustainable development. This will bring a better understanding of sustainability issues into public policy debates as engineers have a greater influence on policy and infrastructure development.

References

Beckerman W (2003) A poverty of reason. The Independent Institute, Oakland

Bronowski J (1973) The ascent of man. Little Brown, Boston

Chadderton R (2010) Does study of economy and risk alter engineers' views of sustainability issues? World Environment & Water Resources Congress, Providence

Committee on Sustainability (2004) Sustainable engineering practice: an introduction. ASCE, Reston

Davidson C, Matthews H, Hendrickson C, Bridges M, Allenby B, Crittenden J, Chen Y, Williams E, Allen D, Murphy C, Austin S (2007) Viewpoint: adding sustainability to the engineer's toolbox: a challenge for engineering educators. J Envr Sci Technol 41(14):4847–4849

Hatch H (2002) Sustainable development, excerpts from an address to the Presidents' Circle of the National Academies on November 14, 2002, Washington DC

Hayek F (1974) Prize lecture: The pretence of knowledge. Nobelprize.org, 17 Sep 2012. http://www.nobelprize.org/nobel_prizes/economics/laureates/1974/hayek-lecture.html

Hayek F (1977) The use of knowledge in society. Institute for Humane Studies, Menlo Park

Mises L (1996) Human action, 4th rev. edn. The Foundation for Economic Education, Irvington-on-Hudson

Sabin P (2013) The bet. Yale University Press, New Haven

Sherman D (2008) Sustainability: what's the big idea? Sustainability 1(3):188–195

Simon J (1981) The ultimate resource. Princeton University Press, Princeton

Solow R (1992) An almost practical step toward sustainability. Resources for the Future, Washington DC

Solow R (1993) Sustainability: an economist's perspective. In: Dorfman R, Dorfman N (eds) Economics of the environment, 3rd edn. Norton, New York

Stroup R (2003) Eco-nomics. Cato, Washington DC

Wildavsky A (1988) Searching for safety. Transaction Publishers, New Brunswick

Wildavsky A (1990) No risk is the highest risk of all. In: Glickman T, Gough M (eds) Readings in risk. Resources for the Future, Washington DC pp 120–127

Wildavsky A (1995) But is it true? Harvard University Press, Cambridge

Appendix A
Elements of Austrian School Economics

The glossary for Mises' *Human Action* that was developed by Greaves (1974) is introduced by the following quotation:

"The first purpose of scientific terminology is to facilitate the analysis of the problems involved." (*Human Action*, p. 434).

Greaves (1974) added that the importance of a definition is not whether it is a popular one, "but rather does it lead the reader to a better understanding of the highly complex ideas the author is trying to elucidate."

The purpose of this Appendix is to identify a number of the elements of Austrian economic theory that should be more widely known, especially by future civil engineers. My intention is to share with instructors the way I try to explain these ideas to undergraduate engineering students who have no background in economics, especially that of the Austrian School.

As justification for including Austrian elements in an engineering economy class, we should note that we are continually hearing from the highest levels of society the very fallacies falsified long ago by the Austrians. One recent example was a statement by a well-known economist that damage by Hurricane Sandy stimulated the economy. I remember a high-ranking member of Congress say the same thing about World War II. This is why I often ask the students whether it is really the work created by such destruction that is highly valued. Were not the physical things that were destroyed more important, not to mention the staggering loss of life?

Austrian economics is non-prescriptive in the sense that it claims to evaluate means and not ends. Austrian economic theory asks whether a means is appropriate to achieve the end sought. There is no concern for the correctness of any actor's preferences. Consequently, income and wealth of individuals are not considered arbitrary but rather reflect how well the individual contributed to satisfying consumer "wants." This means that Austrian economics has no direct moral dimension. Rizzo (1992) concluded that Austrian economic principles are value-free because they do not incorporate the values of the economist making them.

Machlup (quoted by Schulak and Unterkofler 2011) considered that the six most important characteristics of the Austrian School were:

© Springer International Publishing Switzerland 2015
R.A. Chadderton, *Purposeful Engineering Economics*,
DOI 10.1007/978-3-319-18848-5

1. Methodological Individualism.
2. Methodological Subjectivism.
3. Taste and Preferences.
4. Opportunity Costs.
5. Marginalism.
6. Time Structure of Production and Consumption.
    Machlup (Schulak and Unterkofler 2011) added two further characteristics that he associated particularly with the Mises branch of the School:
7. Consumer Sovereignty.
8. Political Individualism.

This level of condensation cannot adequately describe Austrian Economics to a student new to the subject. A more detailed list of topics for the purposes of my course is presented below. Even this cannot be a complete summary of the Austrian School of Economics because that would be virtually impossible in any reasonable time or space. Only the most significant discussions for the purposes of this book and for my course are included.

An invaluable reference is the pamphlet by Taylor (1980) who maintained that "…the Austrian analysis cannot be overlooked if a greater understanding of the market process and the effects of interferences with its operation is to be achieved." I have tried to incorporate the main elements of Austrian economics in class according to the following explanations. Additional references that an instructor might find useful are listed at the end of this Appendix.

Human action is based on methodological individualism. The foundation of all of Austrian Economics is that each individual is a conscious actor, has chosen goals, personal preferences, and attempts to achieve desired ends. Society is seen as merely the complex of interacting individuals, not an independent entity.

Subjective value theory, based on marginal utility, is not quantifiable. This means that each individual assigns a ranking of preferences. Priorities must be assigned because wants always exceed means; resources are always in short supply. Marginal utility is not measurable.

Time preference means that current goods will always be more highly valued than future goods. The future is discounted to the present. Under free conditions this natural rate of discounting would basically set an interest rate.

Conceptual understanding is more reliable than mathematics. Mathematics is "singularly inappropriate" to the science of human action because humans are conscious, causal actors. Consequently, statistics and correlation do not apply to human action and econometrics is a failure.

Resources are always in short supply and the economic problem is to discover the best use of available resources.

The division of knowledge is a problem for the coordination of human action. Hayek's description of the use of knowledge says "the division of knowledge is the central problem of economics as a social science." No one can acquire all the information necessary to create a planned economy. The problem is how best to make use of the knowledge spread around among many individuals who do not know about each other's information. The article "I, Pencil" provides an excellent example.

Our individual knowledge is always incomplete and imperfect. The search for improved knowledge is continuous and that means that in economic action there are no constants such as those required for mathematical equations. Consequently, calculations are speculative and uncertain.

Money is an evolved social institution not a legal tender definition by government. Money prices are exchange ratios between goods and amounts of money. Money prices are not measures of value.

Economic calculations are either prospective or retrospective. Economic calculation requires money prices and cannot be in kind. To provide reliable guidance the money used in calculation must have consistent value itself.

Diminishing marginal utility also applies to money as each additional unit of money is seen as having less value to the holder.

Money prices are not costs. Cost is a foregone opportunity, the next most highly valued thing that must be given up for the thing obtained.

Exchange is always to the mutual benefit of both actors. Each item exchanged must be more highly valued by the receiver than by the giver.

Supply and demand relationships are seen as qualitative, not quantitative. Increasing price results in increasing supply and decreasing demand but these amounts are not numerically determinate. Such aggregated amounts are not numerically measurable.

Sunk costs of past actions are irrelevant for the future because economic action is always looking forward.

Information flow is from consumers to producers. Therefore, consumers ultimately determine prices and allowable costs. In the long term, costs of production do not determine prices.

The market is a process not a thing. It is the innumerable interactions of all the participating economic actors.

Prices are not measures of value. They are old data or exchange values from the recent past. Such prices can act as estimates of future prices and as guides to future action.

An understanding of profit or loss can be enhanced by retrospective calculation which could reveal unsustainable capital consumption or mal-investment.

Inflation, as seen by the Austrians, is a policy of intervention by government. It is not caused by demand or supply variations. The problem is that inflation disrupts the market process by distorting the information transfer function of the money. The resulting distortions are not sustainable in the long term.

Hayek's molasses analogy describes the propagation of rising prices throughout the economy after injection of new money by government. The earlier recipients of the new money benefit from increased purchasing power while the later recipients suffer from higher prices throughout the economy. The apparent benefits of the inflation cannot be sustained without increasing the rate of injection of new money.

The liquidation of mal-investment resulting from an inflation is called depression. It is seen as unavoidable. The only question is whether to continue the inflationary policy in an attempt to prop up mal-investment or to stop and allow the adjustment to occur. Continuing a policy of increasing inflation would ultimately lead to the so-called "crack up boom" as described by Mises.

Austrian theory emphasizes the critical difference between what is seen in the short term and the unseen, long-term consequences of even well-intentioned, collective actions. The long-term consequences for everyone must take precedence over the short-term benefits for any special interest group.

Austrian analysis advocates a free market system because it allows for discovering and correcting error, while central planning and interventionism do not. As seen by Hayek, a free economy would not get stuck or break down. The market coordinates planning by many individuals to promote social efficiency.

Mises said that market capitalism and socialism are the only choices. A mixed economy will eventually degenerate into one or the other; there is no middle way. A free market society is seen as the only possible way to maximize the unmeasurable satisfaction levels of the members of that society.

Suggested References

Bastiat F (2011) That which is seen, and that which is not seen. Dodo Press, UK

Caldwell B, Boehm S (1992) Austrian economics: tensions and new directions. Springer, New York

Chadderton R (1983) Praxeology and engineering. J Prof Issues Egr 109(3):159–169

Chadderton R (1994) Relation of praxeology to engineering. J Prof Issues Engrg Educ Practice 120(4):384–391

Gloria-Palermo S (1999) The evolution of Austrian economics. Routledge, London

Greaves P (1974) Mises made easier. Free Market Books, Dobbs Ferry

Gwartney J, Stroup R, Lee D, Ferrarini T (2010) Common sense economics, revised ed. St. Martin's Press, New York

Hayek F (1977) The use of knowledge in society. Institute for Humane Studies, Menlo Park

Hayek F (1991) The fatal conceit. Chicago

Hayek F (1992) The fortunes of liberalism: essays on Austrian economics and the ideal of freedom. Liberty Fund, Indianapolis

Hayek F (2009) A tiger by the tail. Ludwig von Mises Institute, Auburn

Hayek F (2011) The constitution of liberty. Chicago

Hazlitt H (1996) Economics in one lesson. Laissez Faire Books, San Francisco

Menger C (1994) Principles of economics. Libertarian Press, Grove City

Mises L (1996) Human action, 4th revised edn. The Foundation for Economic Education, Irvington-on-Hudson

Mises L (2005) Liberalism. Liberty Fund, Indianapolis

Mises L (2012) Theory and history. Martino, Mansfield Centre

O'Driscoll G, Rizzo M (1996) The economics of time and ignorance. Routledge, New York

Read L (1992) I, pencil. Imprimis 21(6):1–3

Rizzo M (1992) Afterword: Austrian economics for the twenty-first century. In: Caldwell B, Boehm S (eds) Austrian economics: tensions and new directions. Springer, New York, pp 245–255

Rothbard M (1980) The essential von Mises. Libertarian Press, Grove City

Rothbard M (2004) Man, economy, state. Ludwig von Mises Institute, Auburn

Schulak E, Unterkofler H (2011) The Austrian School of economics: a history of its ideas, ambassadors, & institutions. Ludwig von Mises Institute, Auburn

Sowell T (2000) Basic economics. Basic Books, New York

Taylor T (1980) The fundamentals of Austrian economics. Cato, San Francisco

Vaughn K (1998) Austrian economics in America. University of Cambridge Press, Cambridge

Williams W (2000) The legitimate role of government in a free society. Imprimis 29(8):1–4

Williams W (2008) Liberty versus the tyranny of socialism. Hoover, Stanford

Appendix B
Example Learning Objectives

My course, "Economy and Risk," was presented for the final time in the fall of 2012. There were 28 class meetings of 75 min each and a three-hour final exam period. The list of possible "learning objectives" essentially follows my order of topics; but other engineering economy textbooks might have a different order. These objectives are intended to give an instructor ideas about ways to include some Austrian concepts within an engineering economy class.

Learning Objectives for "Economy and Risk" Fall 2012

To complete this course successfully, each student should be able to satisfy the following 'learning' objectives.

Explain the main points of each chapter so that a high school senior would understand.
Explain the "origin of money" to politicians.
Explain "operating at the margin" as an economic decision rule.
Explain Gresham's Law to journalists.
Explain the "use of knowledge" in society to a liberal arts major.
Distinguish simple from compound interest.
Draw cash flow diagrams for discrete payment series.
Apply discrete-payment, discrete compounding formulas to solve problems.
Calculate P, F, A using interest formulas.
Distinguish nominal and effective interest rates.
Calculate effective interest rates for various compounding intervals.
Explain interest formulas to a liberal arts major.
Explain the concept of "equivalent" cash flows.
Determine whether cash flows are equivalent.
Calculate equivalent amount of a cash flow at a present or a future time.
Calculate the repayment schedule for a loan.

© Springer International Publishing Switzerland 2015
R.A. Chadderton, *Purposeful Engineering Economics*,
DOI 10.1007/978-3-319-18848-5

Calculate the effective interest rate for a repayment schedule.

Explain the meaning of price index, price increase, and inflation to a reporter.

Explain the "molasses analogy" to politicians.

Calculate average inflation rate from a price index.

Distinguish among market-rate, inflation-free rate, and inflation rate.

Convert from market rate to inflation-free rate and vice-versa.

Solve cash flow problems by "actual" and "constant" dollar analyses.

Analyze effect of inflation on an investment (such as, your retirement account!).

Calculate PW, AE, FW, and IRR for a general flow of discrete payments.

Calculate capitalized equivalent for a project.

Calculate capital recovery with return.

Generate a chart of project balance with interest as a function of time.

Compare and select alternatives based on total investment comparisons.

Explain the decision rules studied to a math major.

Explain the concepts of a "decision criterion", MARR, opportunity cost, and the "do nothing alternative" to a journalist.

Apply PW, AE, FW, and IRR to incremental investment analysis problems.

Compare and select alternatives based on incremental analysis.

Choose among alternatives having unequal lives.

Explain break-even analysis to a politician.

Explain economy of scale to a liberal arts major.

Explain sunk cost and opportunity cost to several politicians.

Explain the fundamental process of B/C analysis to a lawyer.

Explain the significance of choosing a proper interest rate for B/C analysis to a politician.

Apply correct B/C analysis to select the best alternative.

Explain the principles of, and need for, project cost accounting to a liberal arts major.

Define direct and indirect cost; direct labor and billable time; overhead; and profit.

Explain the concepts of "depreciation" and "book value".

Explain the effect of accelerated depreciation on project NPV.

Apply SL and accelerated methods to tabulate allowable depreciation vs. time.

Calculate effects of depreciation methods on cash flow NPV.

Explain common effects of interest, depreciation, inflation, and taxes on cash flow to several politicians!

Convert a "before tax" cash flow to "after tax" cash flow while accounting for effects of inflation and depreciation.

Apply the previous decision rules to ATCF.

As a basis for the risk component of the course, students need knowledge of selected topics in probability and statistics. Learning objectives that I have emphasized in my classes have included:

Explain the idea of expected value decision making to a liberal arts major.

Construct a decision tree model.

Estimate the expected value of perfect information for a decision tree model.

Explain the difference between risk and uncertainty to a math major.

Identify and explain the three R's of safety assessments.

Explain the "axioms of probability" to a liberal arts major.

Calculate the union, intersection, and conditional probabilities from a Venn diagram.

Use Total Probability and Bayes' Theorum to solve imperfect testing problems.

Use normal and standard normal distributions to estimate event probabilities.

Explain types of uncertainty to a politician.

Estimate mean and variance from data.

Calculate confidence intervals for the mean.

Construct significance tests for model hypotheses.

Index

© Springer International Publishing Switzerland 2015
R.A. Chadderton, *Purposeful Engineering Economics*,
DOI 10.1007/978-3-319-18848-5

Printed in the United States
By Bookmasters